Gateway to VLSI

Want to be an FPGA Engineer?
Edition 1

Bharat Agarwal
Kshitij Goel

INDIA · SINGAPORE · MALAYSIA

Notion Press

Old No. 38, New No. 6
McNichols Road, Chetpet
Chennai - 600 031

First Published by Notion Press 2019
Copyright © Bharat Agarwal, Kshitij Goel 2019
All Rights Reserved.

ISBN 978-1-64650-812-9

This book has been published with all efforts taken to make the material error-free after the consent of the author. However, the author and the publisher do not assume and hereby disclaim any liability to any party for any loss, damage, or disruption caused by errors or omissions, whether such errors or omissions result from negligence, accident, or any other cause.

While every effort has been made to avoid any mistake or omission, this publication is being sold on the condition and understanding that neither the author nor the publishers or printers would be liable in any manner to any person by reason of any mistake or omission in this publication or for any action taken or omitted to be taken or advice rendered or accepted on the basis of this work. For any defect in printing or binding the publishers will be liable only to replace the defective copy by another copy of this work then available.

Special Thanks

To

Dhiraj Khanna

Director

Pythian Technologies

CONTENTS

Foreword *7*

Preface *9*

Acknowledgement *11*

Organization of the Book *13*

PART I : Verilog 15

PART II : Field Programmable Gate Array 45

PART III: Protocols 83

PART IV: Static Timing Analysis 109

PART V : Important Codes 187

FOREWORD

VLSI design has grown at an exponential rate in conformance with Moore's Law. Complexity has gone from a million to a billion transistors in the last 20 years, and dimensions have seen reductions from 1um to now 7nm. With technology making such rapid strides, companies too are clamouring for acquiring the right talent for furthering their business objectives in this exciting field. Popular job portal sites are awash with jobs for Digital Design Engineers and Design Verification Engineers. For an aspirant, not only is it important to keep abreast with technology but also know how to crack that all-important job interview.

This book aims to assist the aspirant in adopting a focussed approach towards tackling interviews. Having been through the grind themselves, the authors have painstakingly curated this repository of probing questions that are likely to be asked in such interviews. The level of questions varies from those appropriate for freshers as well as the experienced VLSI engineer. I am quite happy to see the depth to which this concise compendium goes. Unlike other sources available, this is perhaps the first text that combines questions ranging over the entire VLSI landscape from Verilog, FPGA, various protocols, Static Timing Analysis as well as important code snippets. This book touches every aspect of the expectations out of a VLSI engineer, and I have no doubts in strongly recommending this book to all budding aspirants.

The authors have spoken to both interviewees and interviewers while compiling this repository over a long period. Being practising engineers themselves, I am really happy to see their labour of love take shape in the form of this book. Their innovative and enthusiastic mindset is reflective in the quality of questions that they have framed and this is a good read for the interviewers too. I wish the authors the very best in their quest for helping all aspirants through this concise and precise curation of relevant information.

– **Sanjib Chaudhary**

CTO, VaaaN

PREFACE

You may be a self-motivated and ambitious engineer. Perhaps you are a student or professional. Regardless of your position or occupation, we have one goal in common: the desire or the need to be at the forefront of the fascinating field of FPGA, Verilog, and Static Timing Analysis.

Are you reading this preface to find out why you should read this book or what you will learn from it? How is this book unique? Should you use it as a reference text in your courses? Our role is to cooperate with you in the learning process. We wish to share our product and system design experience and enthusiasm to improve our joint understanding of FPGA, Verilog, and STA.

This book contains more than 105 interview questions related to the field of FPGA, Verilog, Static Timing Analysis and Protocols. This book provides a platform which comprises of top questions which help them in clearing the interview of the world best companies related to FPGA, Verilog. The protocols include the UART, SPI and I2C serial communication protocols with detailed analysis. This book also provides some important Verilog codes like FIFO etc.

The main aim to write this book is to help professionals and graduates that instead of looking to google for the answer of any question related to FPGA, Verilog, protocols or STA we bring all those questions on a single platform with detailed analysis and real-time examples which help them in nurturing their concepts and clearing the interviews.

It is a real challenge to produce a "high-tech" book within a reasonable time limit in a very rapidly evolving field. Within our constraint of time, we did our best to write an authoritative and comprehensive book. We hope that you will be motivated and challenged and that you may even "fall in love" with this fascinating field as we have throughout the years. For our self, the largest professional gratification we can receive is to see this book contribute to engineering knowledge and achievements and help graduates and professionals to get there dream job.

ACKNOWLEDGEMENT

It is a great pleasure and honour for us to be associated with Notion Press. We want to express our sincere gratitude and thanks to the entire editorial team and production department at Notion Press, New Delhi for publishing the book and maintaining a high degree of precision and accuracy. We want to thank god for providing such a great opportunity in our life. We want to thank our parents, who were the guiding force and where we stand all credit goes to their values and ethics that we learned from them. Writing a book is harder than we thought and more rewarding than we could have ever imagined.

Many supported us during this interesting journey of writing the book, but some are worth to be mentioned. We want to thank Ankit Kumar for giving his valuable inputs in shaping the contents of the book. We both would like to mention our under-graduate professors Mrs.Babita Tyagi who supported us in our niche days and because of their guidance, we are on the zenith of life. We would also like to thank our friends Shivankar Swarup, Rahul Kumar, Sanjit Varma, Konika Pandey, Vipul Khatiyan, Kripa Shankar Sharma and Avinash Swami for giving constant support and motivation.

We would also like to thank our school teacher Mr B.P Singh and Mrs Sonakshi of Vidya Bharti School, who laid a strong foundation in our interesting and challenging journey.

Every effort has been made to produce an error-free text; however, we would be grateful if readers can point out any unintended error or discrepancy. They can feel free to write to us bharatreg52@gmail.com, kshitijgoel92@gmail.com or to Notion Press, New Delhi, to give any suggestions or feedback to enhance the quality of the book.

ORGANIZATION OF THE BOOK

The book is divided into five sections.

PART I: VERILOG

In this section, you will find top questions related to Verilog that can be asked by top companies' interviewers. This section will touch every concept related to Verilog from niche to zenith.

PART II: FPGA

In this section, you will find a detailed analysis of the FPGA from its architecture to its functioning. We have tried to give images in every question to make understanding more intact and clearer.

PART III: PROTOCOLS

In this section, we talk about some famous serial communication protocols like I2C, SPI and UART. You will find the detailed analysis of each protocol with its advantages, disadvantages, waveform, speed and applications.

PART IV: STATIC TIMING ANALYSIS

In this section, you will find real-time examples that you will face when you work on high-end projects. We have provided detailed solutions for every problem with timing waveforms and exact numbers. This section will overall help you in understanding the importance of timing analysis in designing any digital circuit.

PART V: IMPORTANT CODES

We have also provided the Verilog codes for some important topics like FIFO, Sequence Detector. The codes have been written in a very simple way, and enough comments have been mentioned in the codes which will help the users in having better understanding.

PART I

VERILOG

Q1. What is **Stuck at 0** and **Stuck at 1** means in VLSI?

Ans. When a signal, or gate output, is stuck at a 0 or 1 value, independent of the inputs to the circuit, the signal is said to be "stuck at" and the fault model used to describe this type of error is called a "stuck at fault model".

Q2. Explain **'timescale 1ns/1ps** and **'timescale 10ps/1ps**?

Ans. 'timescale <time_unit>/<time_precision>.

time_unit = The time multiple for time values.

time_precision= Minimum Step Size during Simulation.

1ps=.001 ns #1 //= 1 ns delay.

#1.003; //=will be considered as a valid delay.

#1.0009; //= will be taken as 1 ns since it is out of the precision value.

For, 'timescale 10ps/1ps

time_unit is 10 ps and time_precision is 1ps.

For example: nor #3.57(z,x1,x2).

When this command gets simulated, then simulation time will be 36 ps to 3.57 ps because of 3.57ps*10 (time_unit) =35.7 ps and precision is 1 ps, so it becomes 36 ps.

Q3. What are the different types of data types in Verilog?

Ans. Data types in Verilog :

1. Net (physical connectivity) – default value z

    a. wire, wand, wor

    b. tri, triand, trior

    c. tri0, tri1, trireg

    d. supply0 (GND), supply1 VDD (VCC)

2. Registers (physical storage elements) – default value x

   a. reg, integer, real

   b. time, real-time (time is in real number format)

Notes:

1. "wor" performs "or" operation on multiple driver logic.

2. "wand" performs "and" operation on multiple driver logic.

3. "trior" and "triand" perform the same function as "wor" and "wand", but model outputs with resistive loads.

Q4. What is simulation time?

Ans. The simulation time is used to refer to the time value maintained by the simulator to model the actual time it would take for the system description. The term is used interchangeably with simulation time.

Q5. What is a Time Slot?

Ans. A time slot encompasses all simulation activity processed in an event region for each simulation time. All simulation activities for a simulation time got executed until further simulation activity remains for that particular time slot, i.e. without advancing the simulation time.

Note: The execution of simulation events within a time slot may require multiple iterations through the simulation event regions for that same time slot.

The IEEE Std 1800-2005 standard sometimes referred to a time slot as a time step, but the term timestep has removed from the P1800-2008 Draft Standard.

Q6. Difference between Simulation and Synthesis?

Ans. **Simulation** is the process of using simulation software (simulator) to verify the functional correctness of a digital design that is modelled using an HDL (Hardware Description Language),e.g. Verilog and VHDL.

Synthesis is a process in which a design behaviour modelled using an HDL translated into an implementation consisting of logic gates. A synthesis tool, which is another software program performs this process. One main difference in terms of modelling using a language like Verilog is that for synthesis process, the design behaviour should be modelled as an RTL (Register Transfer Level) abstraction which means modelling in terms of the flow of digital signals between hardware registers (flip-flops) and the logical operations (combinational logic) performed on those signals. Hence, if a Verilog design model intended for synthesis process, only synthesizable constructs should be used, while for simulation there are no such restrictions.

Another main difference is the sensitivity list used in always blocks. These are important for a simulator to decide on how to evaluate the logic, but for synthesis, it is a don't care.

Q7. What are non-synthesizable constructs?

Ans. Some of the **non-synthesizable constructs** in Verilog are tasks, delays, events, fork…join, initial…begin blocks, force and release statements. If any code is designed for synthesis process, we should avoid such things.

Q8. Explain the flow of Verilog code in Vivado software?

Ans. The code that is written in Vivado and meant to run on hardware must go three-step process:

1. Synthesis
2. Implementation
3. Bitstream generation

When the bitstream of HDL code gets generated, it gets uploaded on hardware using different methods like USB cable, JTAG etc.

Q9. How is Model-sim software from mentor graphics different from Vivado software from Xilinx?

Ans. The code written in the Model-Sim is meant for simulation purpose only. Model-sim does not provide any access to upload HDL code on FPGA whereas in Vivado we can perform both tasks, we can design code for simulation and HDL code can be uploaded on FPGA as well (if it is synthesizable).

Q10. How is ISE design suite different form Vivado?

Ans. Xilinx has developed both the software. ISE design suite is getting obsolete because the latest version of FPGAs for example 7 series FPGAs cannot programme using ISE design suite, to program 7 series FPGA's Vivado software should be used.

Q11. What are the different timing regions in Verilog?

Ans. Different timing or execution regions are as follows: Active, Inactive, non-blocking active region (NBA), Postponed.

Active Region:

All the blocking assignments are executed, RHS of NBAs are evaluated, continuous assignments, $display command, and inputs and update outputs of primitives evaluated.

Inactive Region:

#0 blocking assignments are evaluated in this region.

Non-Blocking Assign Updates:

Updating of LHS of NBAs is done here.

Postponed Region:

$monitor and $strobe commands are executed here.

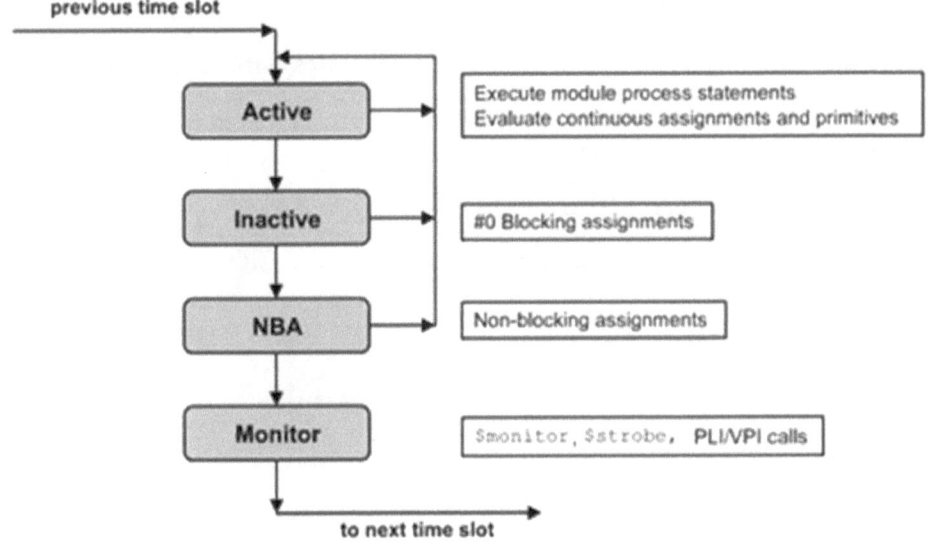

Figure 1.1: Regions of Verilog

Q12. What is the difference between signal and variable in VHDL?

Ans. Signal can be used globally while variable can be used only within the file. Signal took one clock cycle to update while variable doesn't need an extra clock cycle.

Variables need to be defined after the keyword process but before the keyword begin. Signals are defined in the architecture before the begin statement.

Variables are assigned using the: = assignment symbol. Signals are assigned using the <= assignment symbol.

Q13. What are the different types of modelling available in Verilog?

Ans. There are four different types of modelling in Verilog:

1. Gate Level Modelling
2. Data Flow Modelling
3. Behavioural Modelling
4. Switch Level Modelling

Gate Level Modelling

- Hardware implemented in terms of logic gates with interconnections between these gates.
- Design at this level is like describing the design in terms of the gate-level logical diagram (white box).
- It is mainly used to model combinational logic design.

Data Flow Modelling

- Hardware implemented by specifying the data flow.
- A designer is aware of how the data flows between registers(black box).
- It is also used to describe data flow between registers known as RTL (Register Transfer Level) – a type of dataflow modelling used for the synthesis.

Behavioural Modelling

- It is used to model the behaviour of a design without the concern for the hardware implementation details, i.e., at the abstract level or highest level of abstraction.
- Designing at this level is very similar to C programming.
- Mainly used to model sequential logic design.
- It is also used to describe RTL (Register Transfer Level) – a type of behavioural modelling used for the synthesis.

Switch Level Modelling

- The lowest level of abstraction provided by Verilog with switches/transistors having nodes and interconnection between them experienced designers use this style.

Q14. What is the 4-value logic system in Verilog?

Ans. The 4-values on which Verilog language work are:

1. 0
2. 1
3. X (don't care)
4. Z (high impedance)

Q15. What is **vcd** in Verilog?

Ans. **vcd** – stands for **value change dump**.

Verilog system tasks $dumpfile, $dumpvars, $dumpon, $dumpoff, $dumpflush and more are used.

Code for VCD:

initial $dumpfile ("myfile.vcd");

//no arguments, dump all signals in design

initial $dumpvars;

//dump variable in module instance top

//but not signals in modules instantiated under

initial $dumpvars (1, top);

//dump up to 2 levels of hierarchy below top

initial $dumpvars(2, top);

initial begin

$dumpon;//start dumping

#100000 $dumpoff;//stop dumping at time 100,000

end

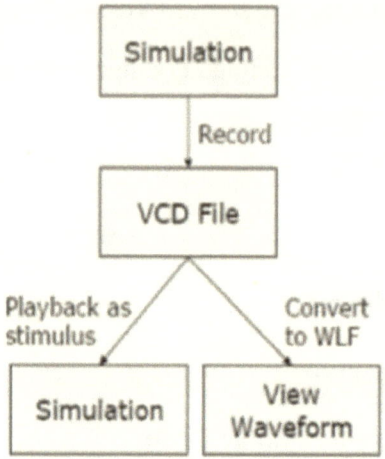

Figure 1.2: Flow chart for VCD

Writing a testbench for vcd dump

module tb_half_adder;

wire s,c;//outputs

reg a,b; // inputs

half_adder ha1 (s,c,a,b);

initial begin

$dumpfile("half_adder.vcd");

$dumpvars(0,tb_half_adder);

end

initial begin

#10 a=1'b0; b=1'b0;

#10 a=1'b0;b=1'b1;

#10 a=1'b1;b=1'b0;

#10 a=1'b1;b=1'b1;

end;

endmodule

A	B	C	D
0	0	0	0
0	1	0	1
1	0	0	1
1	1	1	0

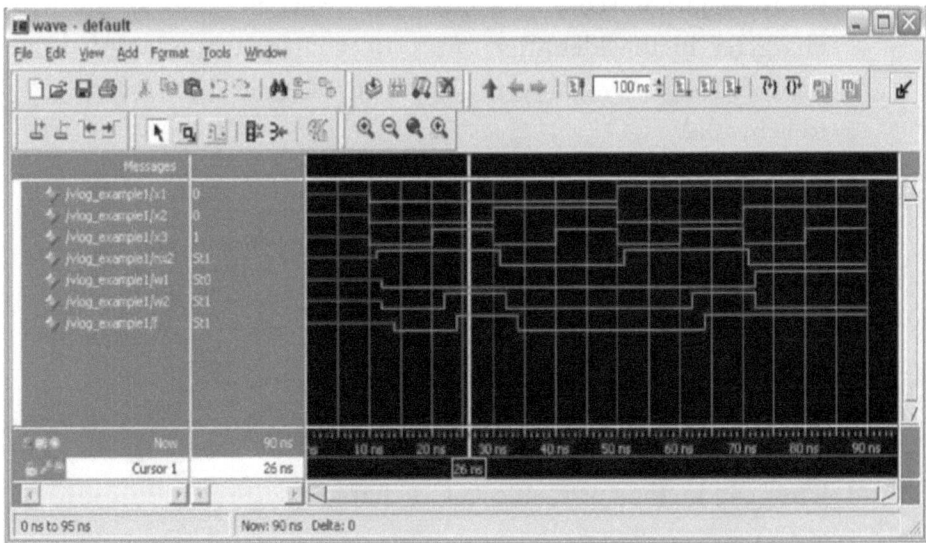

Figure 1.3: Waveform for half_adder in Model Sim

Q16. What is the basic lexical convention used in Verilog HDL?

Ans. The basic lexical convention used by Verilog HDL are like those in C programming. All keywords must be in the **LOWER case,** i.e. the language is case sensitive. White spaces make the code more readable but are ignored by the compiler. Blank space(\b), tabs(\t), newline(\n) are ignored by the compiler. White spaces are not ignored by the compiler in strings.

Comments // single line comment style

/* multi line comment style */

Note: Nesting of comments not allowed

Q17. Explain the naming convention in Verilog?

Ans. Here, are the four important points for naming convention.

1. Objects are given different names; names are nothing but identifiers.

2. Identifiers can be built using a combination of letters [A-Z], [a-z], digits [0-9] (can't be used as the first character in the identifier), underscore _, $ character (can't be used as the first character in the identifier), max 1024 characters allowed in an identifier. e.g. myid, m_y_id, _myid, myid3 are valid, $myid, 3my_id is invalid.

3. White spaces not allowed.

4. Statements are terminated by a semicolon ;

Q18. What are escaped identifiers?

Ans. Below is the escaped identifiers.

1. They start with a \ and end with white space.

2. They can include printable ASCII characters.

3. E.g. \ 546,\.*.&, \ {***}, \ a+b-c, \Gate#3

Q19. Explain the value format in Verilog?

Ans. Five important points before defining any value in Verilog

1. Format for defining a value in Verilog is <size>'<radix><value>.

2. When the size is not specified default, value is 32 bits, e.g. 'bz, 'h9.

3. When radix is not specified default type decimal is taken, e.g. 3.

4. 2'b10, 3'd6, 6'o57, 3'O4, 8'H2d, 32'haA19, 5'B110x0, 6'ozz, 12'hZXb.

5. For negative number use – sign e.g. -6'd3, -3'b11. 6.

Note: Underscore can be used to enhance readability, e.g. 12'o07_24, 12'b000_111_010_100

Q20. Difference between blocking and non-blocking assignments?

Ans. Four major differences between blocking and non-blocking assignments are:

1. A blocking statement will not block the execution that is in a parallel block, means it will execute sequentially while non-blocking assignment allows scheduling of assignments that are executed in a sequential block.

2. A blocking statement is a one-step process, i.e. evaluate the RHS of the expression and update the LHS without any delay while non-blocking is a two-step process, i.e. Evaluate the RHS expression at the beginning of time step and next update the LHS at the end of the time step.

3. Blocking statement is executed in the active region of Verilog stratified event queue while evaluation of RHS of non-blocking statement occurred in an active region and updating of LHS side happened in NBA region.

4. Blocking statement can be used in initial, always blocks and assign statements while non-blocking can only be used in initial and always block but assign statements can not be used.

Blocking Assignment

initial begin

a = 1;

b = 0;

a = b; // a = 0;

b = a; // b = 0;

end

Non- Blocking Assignment

initial begin

a = 1;

b = 0;

a <= b; // a = 0;

b <= a; // b = 1;

end

Q21. What is inferred latch and how it is created?

Ans. A latch is inferred within a combinatorial block where the net is not assigned to a known value. Assign a net to itself will still infer a latch. Latches can also be inferred by missing signals form a sensitivity list and feedback loops.

The proper way of inferring an intended latch in Verilog:

/* Verilog */

always @*

begin

if (en)

q = d;

end

Ways latches are accidentally inferred: Signal(s) missing for the sensitivity list (this is why @* should be used): always @ (a or b) // inferred latch:: "c" missing for the sensitivity list.

begin

out = a + b + c;

end *Missing Condition: always @**

begin

```verilog
case(in[1:0])
2'b00: out = 1'b0;
2'b01: out = 1'b1;
2'b10: out = 1'b1;
// inferred latch "out": missing condition 2'b11/default
endcase
end
always @*
begin
next0 = flop0;
next1 = flop1;
// inferred latch "next2": missing initial condition
next3 = flop3;
case(a[2:0])
3'b001: next0 = in;
3'b010: if(b) next1 = in;
3'b100: if(c) next2 = in;
default: if(!b&&!c) next3 = in;
endcase
end
```

Q22. What is the difference between case, casex and casez in Verilog?

Ans. The basic differences are as follows:

1. Case statement considers X or Z as it is. So, a case expression containing x or z will only match a case item containing x or x at the corresponding bit positions. If no case item matches then default item is executed.

2. Case z statement consider z as don't care.

3. Case x statement consider z and x as don't care.

4. Case statement does not consider the items containing x or z for synthesis. All case statements are synthesizable. A common misconception is "?" does mean a don't care, but it does not. It just another representation of high impedance 'z'.

Example for Case:

If no case item matches, then default item is executed. It more like pattern matching.

case(sel)

00: y=a;

01: y=b;

x0: y=c;

1x: y=d;

z0: y=e;

1?:y=f;

default:y=g;

endcase

Sel	y	Case Item
00	a	00
11	g	default
xx	g	default
x0	c	x0
1z	f	1?
z1	g	default

Example for casez:

Casez statement treat z as don't care.

case(sel)

00: y=a;

01: y=b;

x0: y=c;

1x: y=d;

z0: y=e;

1?:y=f;

default:y=g;

endcase

Sel	y	Case Item
00	a	00
11	f	1?
xx	g	default
x0	c	x0
1z	d	1x
z1	b	01

Example for case x:

Case x treat x and z as don't care.

case(sel)

00: y=a;

01: y=b;

x0: y=c;

1x: y=d;

z0: y=e;

1?:y=f;

default:y=g;

endcase

Sel	y	Case Item
00	a	00
11	d	1x
xx	a	00
x0	a	00
1z	c	x0
z1	b	01

Q23. What are the differences between function and task?

Ans. Task and function are used to break up large procedures into smaller, which helps to make life easier for developing and maintaining Verilog code. In this way, common procedures need to be written only once and execute from different places. Both task and function are called from always or initial block and contain only behavioural statements. Definition of task and function must be in a module. The highlighting difference between task and function is that only they can handle event, delay or timing control statements and function execution in zero simulation time.

Task: Task is declared using task and end task keyword. The task must be used if delay, event or timing control constructs are required in the procedure. It may have zero or more than one input, output an in-out argument. As it executes in non-zero simulation time, it can enable other tasks and functions. It does not return value like function; instead, it can pass multiple values through o/p or in-out arguments.

module mux4x1_using_task(Q,IN,SEL);

input [3:0]IN ;

```
input [1:0]SEL;
output reg Q;
always @ (IN or SEL)
mux(IN,SEL,Q);
task mux;
input [3:0]in;
input [1:0]sel;
output out;
case(sel)
2'b00:out<=in[0];
2'b01:out<=in[1];
2'b10:out<=in[2];
2'b11:out<=in[3];
endcase
endtask
endmodule
```

Function: Functions are declared and terminated using function and endfunction keywords. Function can be used if there is no requirement for specifying delays timing control constructs or events. There is at least one input argument is needed to use function. Function will return a single value and it usually uses for calculations. Also, not that it cannot have o/p or in-out arguments. As it cannot handle timing control statements, delay etc. and executes in "0" simulation time it can only enable another function and not task.

```
module mux4x1_using_function(Q,IN,SEL);
input [3:0]IN ;
input [1:0]SEL;
```

output reg Q;

always @ (IN or SEL)

mux(IN,SEL);

function mux;

input [3:0]in;

input [1:0]sel;

case(sel)

2'b00:mux<=in[0];

2'b01:mux<=in[1];

2'b10:mux<=in[2];

2'b11:mux<=in[3];

endcase

endfunction

endmodule

Q24. What is genvar in Verilog?

Ans. A genvar is a variable used in generate-for loop. It stores positive integer values. It differs from other Verilog variables in that it can be assigned values and changed during compilation and elaboration time. The genvar must be declared within the module where it is used, but it can be declared either inside or outside of the generate loop. Example:

generate

genvar i;

for (i = 0; i< 10; i = i + 1)

begin : gen1

genvar j;

```
for (j = i; j >= 1; j = j - 1)
begin : gen2
reg [0:i] R;
initial
begin
R = i;
$display("%m", R);
end
end
end
endgenerate
```

Q25. What is the difference between begin-end and fork-join?

Ans. Assignment statements within an always block can be put between

1. Begin and end execute sequentially – statement order matters (synthesizable).

2. Fork and join execute concurrently –statement order doesn't matter(not synthesizable).

Q26. What are the different types of FSM?

Ans. A finite-state machine (FSM) or finite-state automaton (FSA, plural: automata), finite automaton, or simply a state machine, is a mathematical model of computation. It is an abstract machine that can be in exactly one of a finite number of states at any given time. The FSM can change from one state to another in response to some external inputs; the change from one state to another is called a transition. An FSM is defined by a list of its states, its initial state, and the conditions for each transition. There are two types of Finite state machine:

1. **Mealy Machine:** Mealy Machine are also finite state machines with output value, and its output depends on the present state and existing input symbol. The length of output for a mealy machine is equal to the length of the input.

2. **Moore Machine:** Moore Machine are finite state machine with output value, and its output values depend only on the present state. The length of output for a Moore machine is greater than input by1.

Q27. What is ODDR?

Ans. 7 series FPGAs have dedicated registers in the OLOGIC to implement output DDR(Double Data Rate) registers. This feature is accessed when instantiating the ODDR primitive. DDR multiplexing is automatic when using OLOGIC. No manual control of the mux-select is needed. This control is generated from the clock. There is only one clock input to the ODDR primitive. Falling edge data is clocked by a locally inverted version of the input clock. All clocks feeding into the I/O tile are fully multiplexed, i.e., there is no clock sharing between the ILOGIC or the OLOGIC blocks. The ODDR primitive supports the two modes of operation:

1. OPPOSITE_EDGE mode

2. SAME_EDGE mode

Q28. Explain different modes of operation(with Waveform) of ODDR?

Ans. **OPPOSITE_EDGE Mode:** In OPPOSITE_EDGE mode, both the edges of the clock (CLOCK) are used to capture the data from the FPGA logic at twice the throughput. This structure is like the Virtex-6 FPGA implementation. Both outputs are presented to the data input or 3-state control input of the IOB. The timing diagram of the output DDR using the OPPOSITE_EDGE mode is shown below:

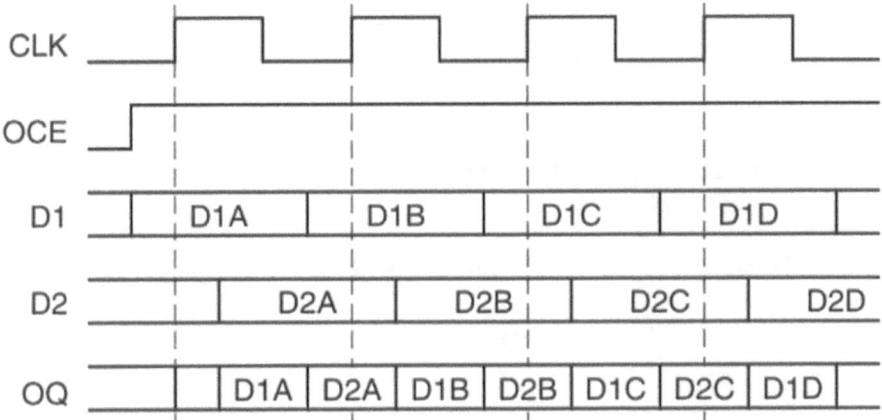

Figure 1.4: Output DDR Timing in OPPOSITE_EDGE Mode

SAME_EDGE Mode: In SAME_EDGE mode, data can be presented to the IOB on the same clock edge Presenting the data to the IOB on the same clock edge avoids setup time violations and allows the user to perform higher DDR frequency with minimal register to register delay, as opposed to using the CLB registers. The timing diagram of the output DDR using the OPPOSITE_EDGE mode is shown below:

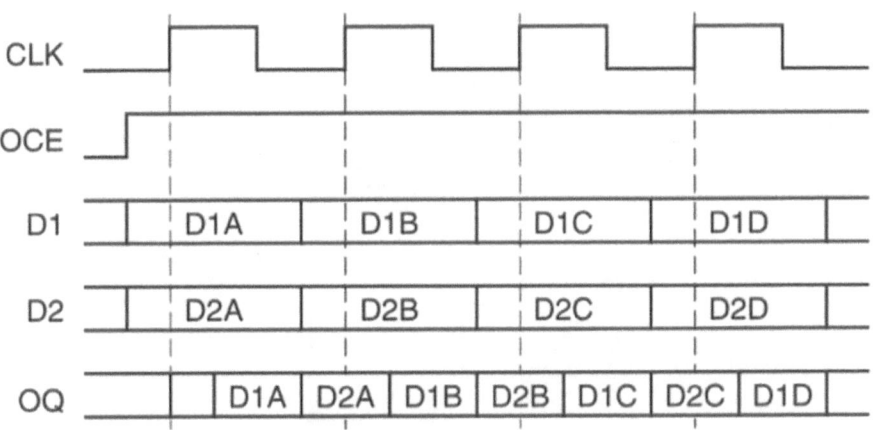

Figure 1.5: Output Timing in SAME_EDGE Mode

Q29. Write the primitive of ODDR in Verilog?

Ans. Primitive of ODDR in Verilog:

ODDR #(

    .DDR_CLK_EDGE("OPPOSITE_EDGE"), // "OPPOSITE_EDGE" OR "SAME_EDGE"

    .INIT(1'b0),   // "Initial Value of Q : 1'b0 or 1b'1"

    .SRTYPE("SYNC") // Set/Reset type: "SYNC" or "ASYNC"

) ODDR_inst (

    .Q(Q),   // 1-bit DDR output

    .C(C),   // 1-bit clock input

    .CE(1'b1),   // 1-bit clock enable input

    .D1(1'b1),   // 1-bit data input (positive edge)

    .D2(1'b0),   // 1-bit data input (negative edge)

    .R(1'b0),   // 1-bit reset

    .S(1'b0)   // 1-bit set

);

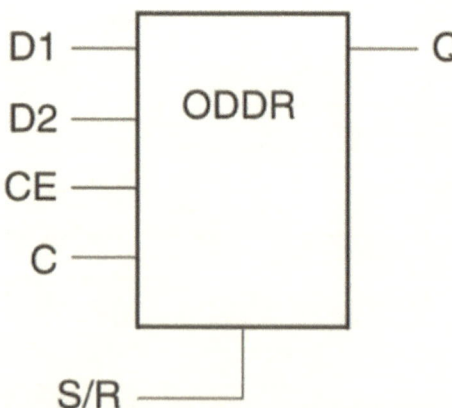

Figure 1.6: ODDR Primitive Block Diagram

Port Name	Function	Description
Q	Data Output (DDR)	ODDR Register output
C	Clock input port	The CLK pin represents the clock input pin.
CE	Clock enable port	CE represents the clock enable pin. When asserted low, this port disables the output clock on port Q.
D1 & D2	Data Inputs	ODDR register inputs
S/R	Set/Reset	Synchronous/Asynchronous set/reset pin. Set/Reset is asserted High.

Note: The ODDR primitive contains both set and reset pins. However, only one can be used per ODDR. As a result, S/R is described instead of a separate set and reset pins.

Q30. What is IDDR?

Ans. 7 series FPGAs have dedicated registers in the ILOGIC blocks to implement inputDDR (Double Data Rate) registers. This feature is used by instantiating the IDDR primitive. All clocks feeding into the I/O tile are fully multiplexed, i.e., there is no clock sharing between ILOGIC and OLOGIC blocks. The IDDR primitive supports the three modes of operation:

1. OPPOSITE_EDGE mode

2. SAME_EDGE mode

3. SAME_EDGE_PIPELINED mode

Q31. Explain the different modes of operation(with Waveform) of IDDR?

Ans. **OPPOSITE_EDGE Mode:** A traditional input DDR solution, or OPPOSITE_EDGE mode, is accomplished via a single input in the ILOGIC block. The data is presented to the FPGA logic via the output Q1 on the rising edge of the clock and via the output Q2 on the falling edge of the clock. This structure is like the Virtex-6 FPGA implementation. The figure below shows the timing diagram of the input DDR using the OPPOSITE_EDGE mode.

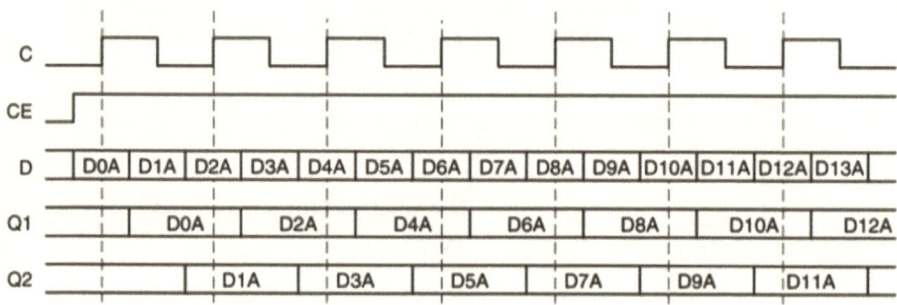

Figure 1.7: Input DDR Timing in OPPOSITE_EDGE Mode

SAME_EDGE Mode: In the SAME_EDGE mode, the data is presented into the FPGA logic on the same clock edge. This structure is similar to the Virtex-6 FPGA implementation. The figure below shows the timing diagram of the input DDR using SAME_EDGE mode. In the timing diagram, the output pairs Q1 and Q2 are no longer (0) and (1). Instead, the first pair presented is pair Q1 (0) and Q2 (don't care), followed by pair (1) and (2) on the next clock cycle.

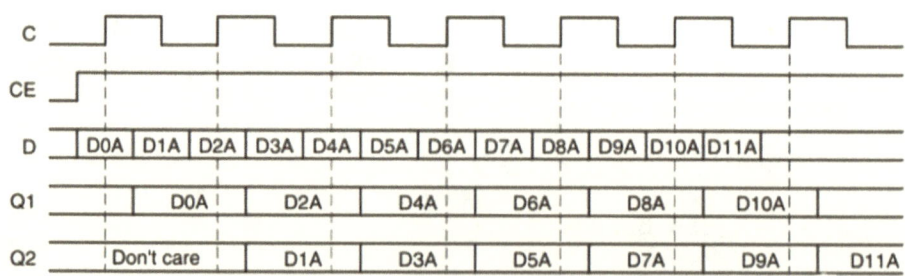

Figure 1.8: Input DDR in SAME_EDGE Mode

SAME_EDGE_PIPELINED Mode: In the SAME_EDGE_PIPELINED mode, the data is presented into the FPGA logic on the same clock edge. Unlike the SAME_EDGE mode, the data pair is not separated by one clock cycle. However, an additional clock latency is required to remove the separated effect of the SAME_EDGE mode. The figure below shows the timing diagram of the input DDR using the SAME_EDGE_PIPELINED mode. The output pairs Q1 and Q2 are presented to the FPGA logic at the same time

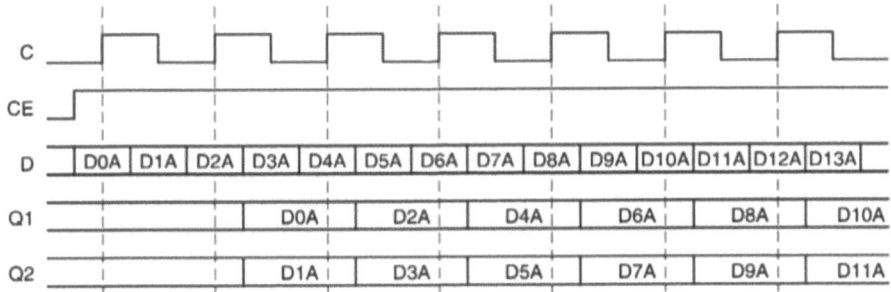

Input 1.9: Input DDR Timing in SAME_EDGE_PIPELINED Mode

Q32. Write the primitive of IDDR in Verilog?

Ans. Primitive of IDDR in verilog:

IDDR #(
    .DDR_CLK_EDGE("OPPOSITE_EDGE"), // "OPPOSITE_EDGE" OR "SAME_EDGE"
    .INIT_Q1(1'b0),   // "Initial Value of Q1 : 1'b0 or 1b'1"
    .INIT_Q2(1'b0),   // "Initial Value of Q2 : 1'b0 or 1b'1"
    .SRTYPE("SYNC")   // Set/Reset type: "SYNC" or "ASYNC"
) IDDR_inst (
    .Q1(Q1),   // 1-bit output for positive edge clock
    .Q2(Q2),   // 1-bit output for negative edge clock
    .C(C),   // 1-bit clock input
    .CE(CE),   // 1-bit clock enable input
    .D(D),   // 1-bit DDR data input
    .R(R),   // 1-bit reset
    .S(S)   // 1-bit set
);

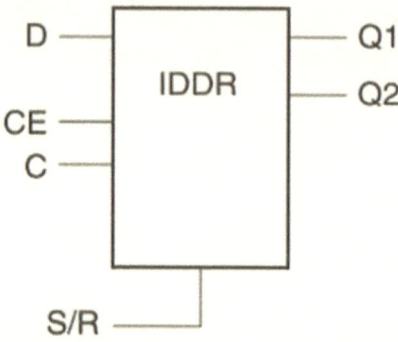

Figure 1.10: IDDR Primitive Block Diagram

Port Name	Function	Description
Q1 & Q2	Data Outputs	IDDR Register output
C	Clock input port	The C pin represents the clock input pin.
CE	Clock enable port	The clock enables pin affects the loading of data into the DDR flip-flop. When Low, clock transitions are ignored, and new data is loaded into the DDR flip-flop CE must be High to load the new data into the DDR flip-flop.
D	Data Input (DDR)	IDDR register input from IOB.
S/R	Set/Reset	Synchronous/Asynchronous set/reset pin. S/R is asserted High.

Note: The IDDR primitive contains both set and reset pins. However, only one can be used per IDDR. As a result, S/R is described instead of a separate set and reset pins.

Q33. Difference between $display, $strobe and $monitor?

Ans: These commands have the same syntax, and display text on the screen during simulation. They are much less convenient than waveform display tools like cwaves.

$display and $strobe display once every time they are executed, whereas $monitor displays every time one of its parameter's changes. The difference between $display and $strobe is that $strobe displays the parameters at the very end of the current simulation time unit rather than exactly where it is executed.

Q34. What is the hardware logic difference in if-else and case statement in FPGA?

Ans. The if-else-if construct infers a priority routing network

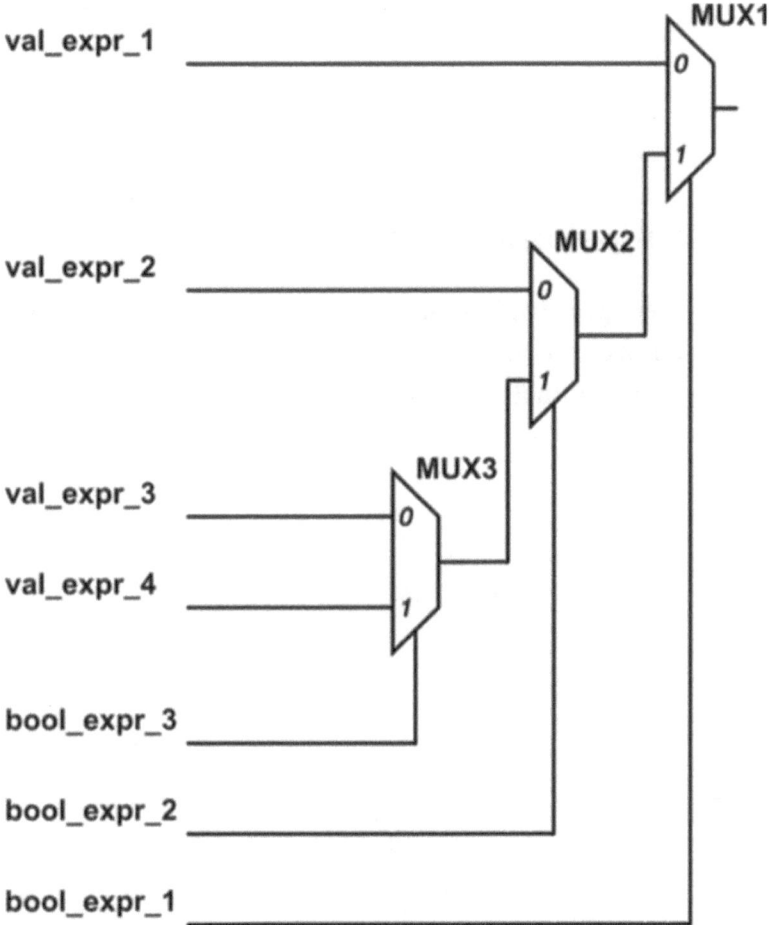

Figure1.11: Hardware Design of if-else-if construct

The case construct, on the other hand, infers a big mux:

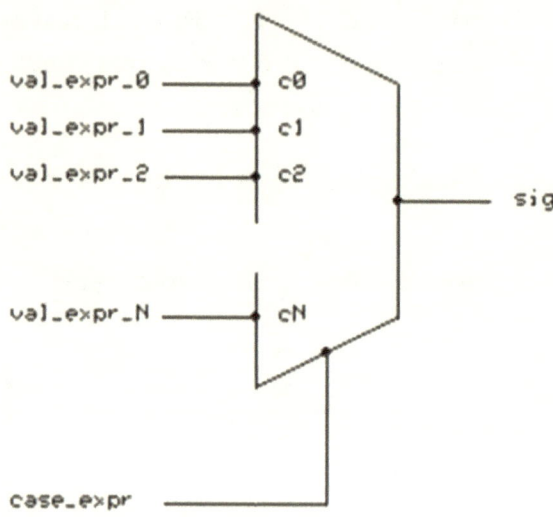

Figure 1.12: Hardware Design of case construct

Since if-else infers priority, it should be used when more than one input condition could occur. Using the case, on the other hand, is appropriate when the inputs are mutually exclusive.

PART II

FIELD PROGRAMMABLE GATE ARRAY

Q1. What is an FPGA?

Ans. Field-programmable gate arrays (FPGAs) are reprogrammable integrated circuits that contain an array of programmable logic blocks. FPGA chip adoption is driven by their flexibility, hardware-timed speed and reliability, and parallelism. Unlike processors, FPGAs are truly parallel in nature, so different processing operations do not have to compete for the same resources. Each independent processing task is assigned to a dedicated section of the chip and can function autonomously without any influence from other logic blocks. As a result, the performance of one part of the application is not affected when you add more processing.

One of the benefits of FPGAs over processor-based systems is that the application logic is implemented in hardware circuits rather than executing on top of an OS, drivers, and application software.

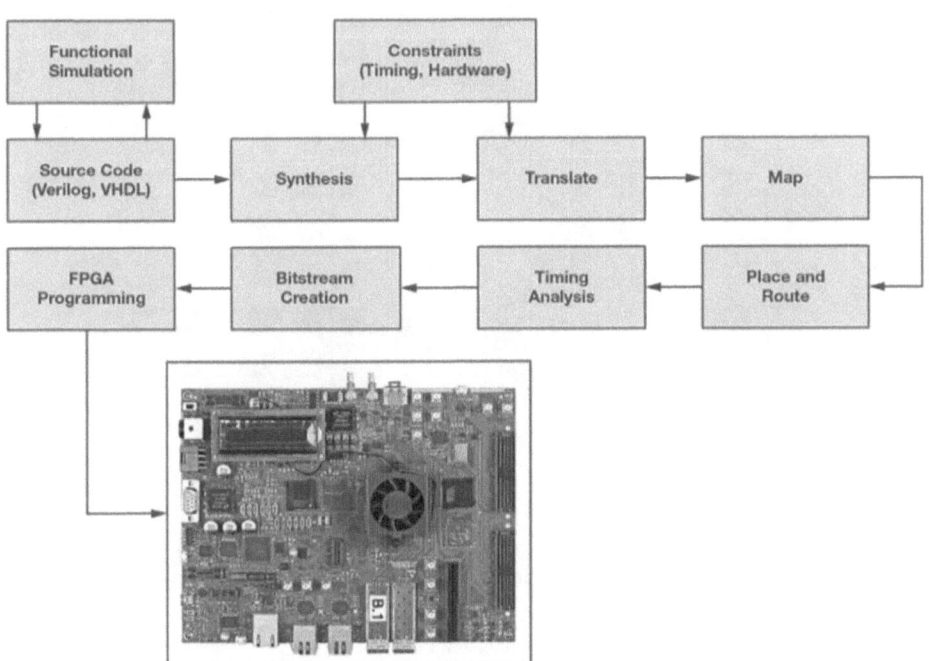

Figure 2.1: Field Programmable Gate Array

Q2. What are different logic resources in FPGA?

Ans. The different logic resources of FPGA are:

1. Flip Flop's
2. LUT's
3. Multiplier and DSP Slices
4. Block Ram (BRAM)
5. Programmable Interconnect
6. I/O Blocks

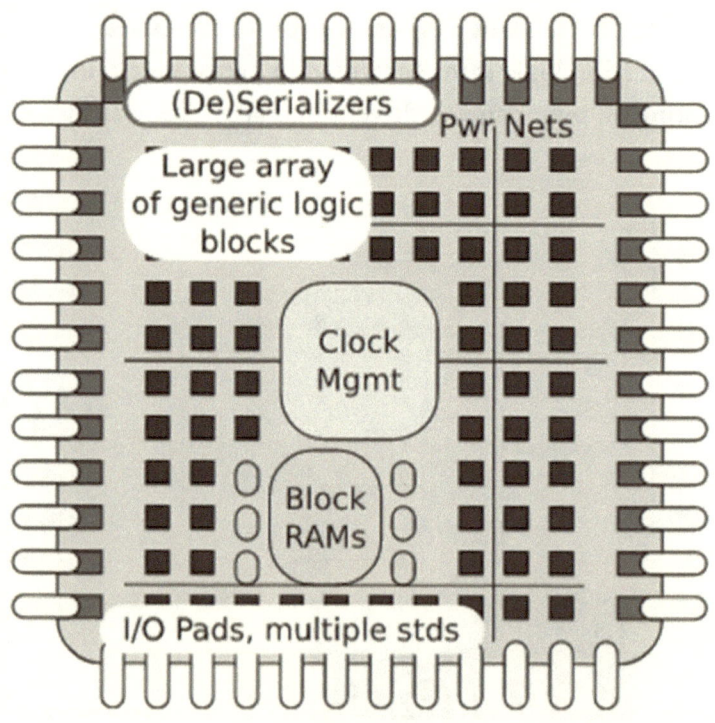

Figure 2.2: Logic Resources insider FPGA

Q3. Difference between CPLD and FPGA?

Ans. CPLD VS. FPGA: NINE KEY DIFFERENCES

1. **Starting the Work:** The CPLD is on the moment you power up the device.

   On the contrary, the FPGA needs some time to get up and to run. It loads the configurations from the external ROM so the delay may take up to several dozen milliseconds.

2. **Stability:** CPLD architecture is persistent. It remains programmed and set up even if the circuit is shut down.

   The FPGA uses SRAM storage of configurations. So, the memory is cleared immediately after the device is turned off.

3. **Timing Analysis:** The number of interconnects used by the CPLD is not so great. The architecture is simple, so timing analysis is fast and seamless.

   FPGA logic is complex. It is rather heavy in comparison with CPLD. Signal routing is not determined, so timing analysis is harder. Some manufacturers provide implementation tools, so this can be fixed. But additional steps are required to improve this factor.

4. **Logic Resources:** The number of logic resources provided by CPLD is relatively small.

   FPGA has myriad logic and storage elements. With them, you can design a complex circuit.

5. **Flexibility:** The fabric scheme does not provide you with enough flexibility. Also, additional components for this purpose are not available.

   With FPGA, you can use many other additional hardware components like blocks, controllers, transceivers, etc., which make the device highly flexible.

6. **Security:** The built-in data storage makes the CPLD tools more secure.

   The FPGA uses external memory, so it is not as safe. There are solutions to handle this, for example, data encryption, Specific security extensions and approaches.

7. **Reprogramming:** To modify the CPLD functionality, you need to turn off the device.

   The FPGA does things differently. It works according to the Partial Reconfiguration method. You can change the circuit on the go: the system runs the design and updates it simultaneously. This feature is especially useful in accelerated programming.

8. **Power Consumption:** CPLD designs don't require too much energy. For example, the newest solutions like CoolRunner-II need only 50uA in ideal conditions.

   On the other hand, we have FPGA which consumes much more energy.

9. **Price:** The CPLD has a rather simple circuit. Therefore, they are quite cheap.

   The FPGA provides more powerful functionality and performance. So, the prices are higher.

Q4. What is CLB in FPGA?

Ans. A CLB (Configuration Logic Block) is a fundamental component of an FPGA, allowing the user to implement virtually any logical functionality within the chip which can be achieved by the usage of two sets of similar components within a block, known as slices. There are two different types of slices, referred to as SLICEM and SLICEL, and each CLB can contain either a SLICEM and SLICEL or two SLICELs. There are almost twice as many SLICELs than SLICEMs on a chip, though the exact numbers vary by device. These slices contain four look-up-tables (LUTs), eight flip-flops (FF), a network of carry logic, and three types of multiplexers.

And to itemize: one CLB = 2 slice, one slice = 4 LUTs + 8 FF.

Therefore, one CLB = 8 LUTs + 16FF.

This numerical summary is to provide an accurate picture of what the contents of a CLB specifically are. These figures may vary from device to device.

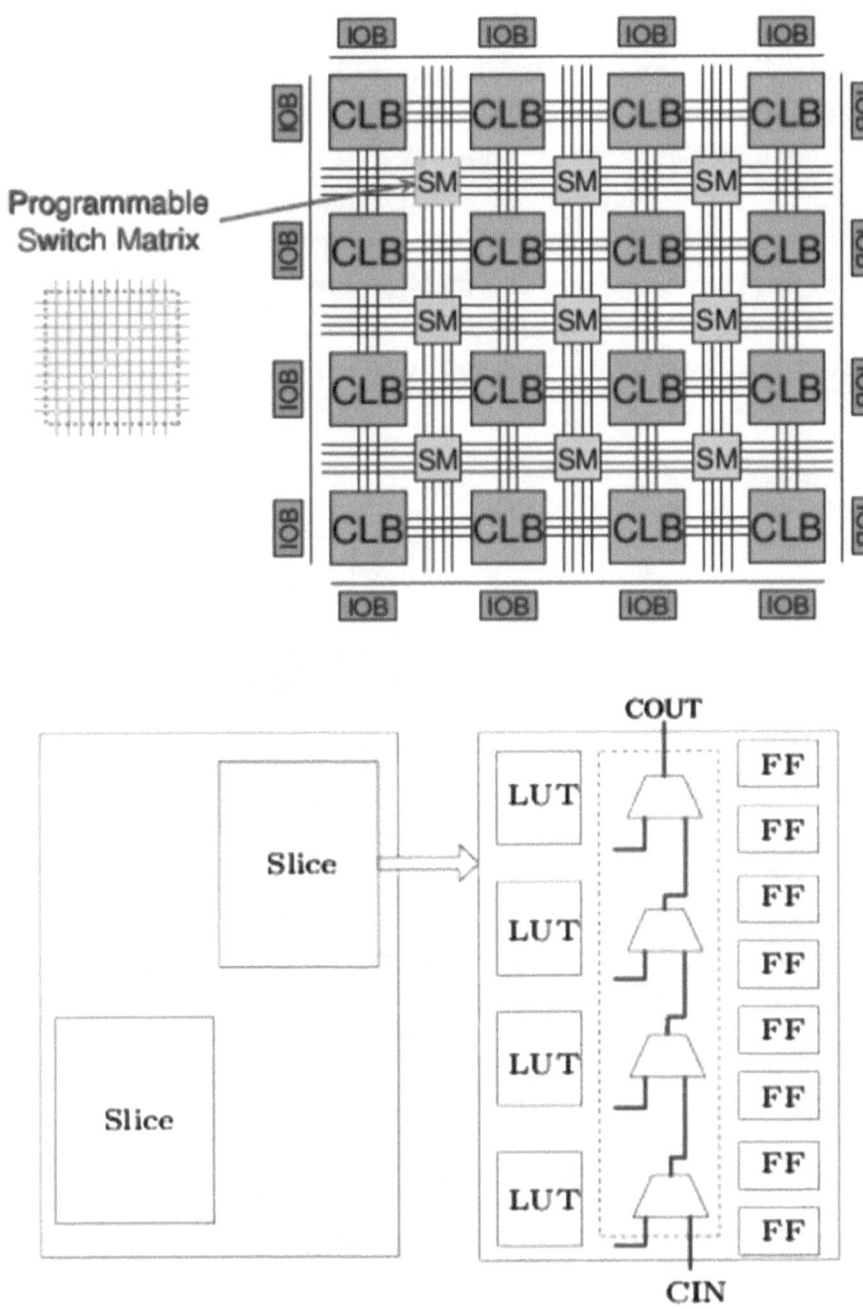

Figure 2.3: Internal architecture of Combinational Logic Blocks

Q5. What are SLICEL and SLICEM?

Ans. The distinguishing feature of the two-slice types is the configuration difference of the SLICEM. SLICEM can be configured so that the look-up tables within it can act as shift registers or as data storage (creating distributed memory on the chip) in addition to its normal logic functionality. A note on naming: the 'M' may be an indication of its ability to act as distributed memory, while the 'L' may be an indication of its exclusive logic functionality. This is just speculative, but it can be helpful to remember which is.

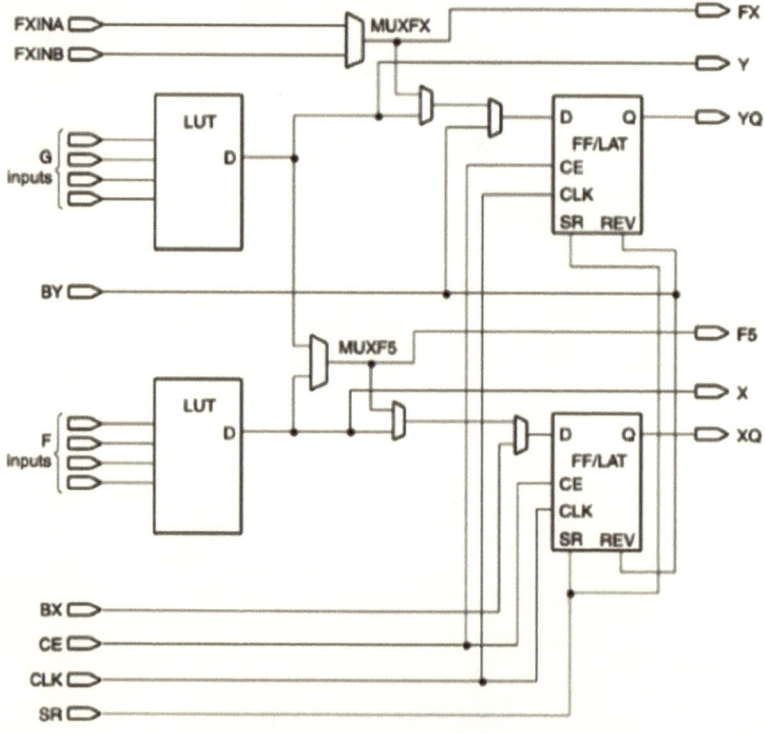

Figure 2.4: Internal architecture of Slice

Q6. What are LUTs in FPGA?

Ans. Look-up tables, the cheat sheet of computing. Okay, this statement may be a bit strong. It does, however, illustrate the point that a LUT can serve to speed up processing by providing a set output for a given input, rather than requiring computation on the input data. The LUTs in an FPGA (Xilinx 7 series) are designed with six inputs and two outputs. Each of these inputs and outputs is independent and allow for the LUT to be implemented in several ways, such as a single six-input function or two functions of five or fewer inputs, though five and four input functions will need to have a common input, while three and two input functions do not. Confusing? Yes. Illustration? Yes!

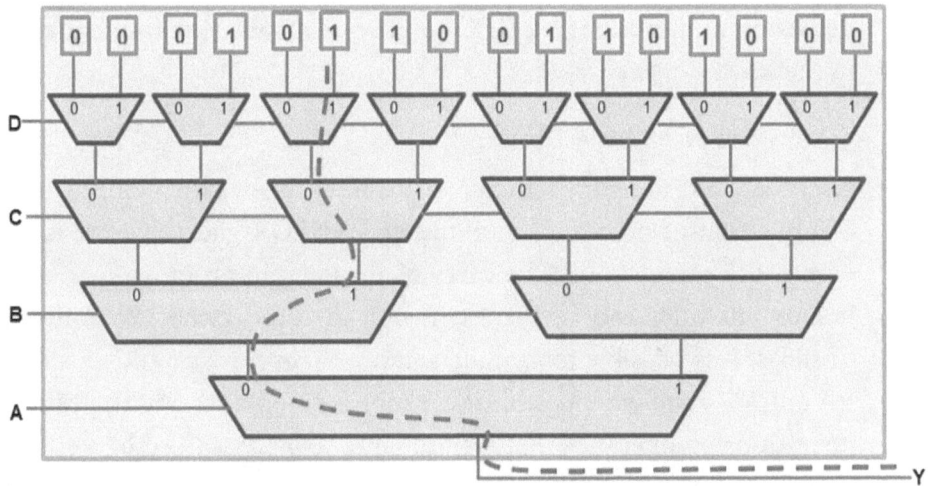

Figure 2.5: Internal architecture of LUT's

Q7. What are Flip-Flop or latches in FPGA?

Ans. A flip-flop is a primitive storage device that can store a single bit of information. Each slice contains eight of these storage elements. Four are available exclusively as flip-flops (synchronous storage), and the other four can be configured either as a standard flip-flop or as a latch (asynchronous storage). One final caveat is that when the four that can be configured as a latch are, the other four flip-flops in the slice become unusable.

Q8. What are Multiplexers in FPGA?

Ans. Multiplexing within a single slice is handled interestingly. Instead of many dedicated multiplexers with fixed inputs, the flexible functionality of the LUT is used. Each LUT can be implemented as a 4:1 Mux, using two of the six inputs as selects for the other four. Larger multiplexing functionality is developed from this starting point. Two LUTs configured in this way are combined into a dedicated multiplexer (called F7AMUX or F7BMUX if the naming is significant to you) which allows you to have an 8:1 MUX. All four LUTs, through the two 8:1 multiplexer, can be combined into a second type of dedicated multiplexer (F8MUX) to provide a 16:1 Mux functionality in the slice.

Q9. What is Carry Logic in FPGA?

Ans. Carry lookahead logic is implemented in each slice on a CLB with a combination of dedicated multiplexer and XOR gates that are used within the carry chain. The carry chain logic in an individual slice is four bits high and is the circuit that directly connects a column of slices. It is possible to cascade a carry chain across multiple slices to quickly implement addition, subtraction, and multiplication operations on operands that are too large to be processed by a single slice (like if you wanted to add/subtract/multiply two numbers bigger than 15 or 1111 for binary speakers).

Q10. What is BRAM in FPGA?

Ans. Block RAM (BRAM) is a type of random-access memory embedded throughout an FPGA for data storage. You can use BRAM to accomplish the following tasks:

1. Transfer data between multiple clock domains by using local FIFOs.

2. Transfer data between an FPGA target and a host processor by using a DMA FIFO.

3. Transfer data between FPGA targets by using a peer-to-peer FIFO.

4. Store large data sets on an FPGA target more efficiently than RAM built from look-up tables.

Q11. What are the different types of data storage and transfer on an FPGA?

Ans. You can store and transfer data on an FPGA using FIFOs, memory items, FPGA registers, or handshake items.

   1. **FIFO:** Use a FIFO when you need to store multiple sequential data samples. FIFOs support lossless data transfer.

   2. **Memory Item:** Use a memory item when you need to access data without respect to the sequence in which the data is written to memory. Unlike FIFOs, all units of data within a memory item are readable at any time. You can also store data in memory items before run time, called pre-initializing data, to use the memory as a read-only reference. Pre-initialized data can save logic resources and computation time.

   3. **FPGA Register:** Use a register when you need to hold only one unit of the specified data size at one time. Data transfer using registers is often lossy, but registers consume fewer FPGA resources than FIFOs when storing equivalent sizes of data. Use registers for synchronisation, and pipelining. **4.Handshake Item**: Use a handshake item when you want lossless data transfer between multiple clock domains inside the FPGA.

Q12. Difference between BRAM and DRAM?

Ans. **DRAM** (Dynamic Random Access Memory): Read and write memory, but needs to be regularly refreshed during which you can't access the cells being refreshed. Off-chip RAM (DRAM) has high bandwidth, but that bandwidth must be shared between all the blocks. Latency is high and can be extremely high if too many blocks are trying to access the RAM at once. On the other hand, it's not at all unusual to find a board with 1GB of DDR3 RAM fitted - enough for thousands of images. Because the DRAM is shared, you can easily use it to transfer data between blocks (and this is one of the main ways of getting data between the PL and PS).

BRAM: SRAM built into the FPGA. Block RAM has high bandwidth and extremely low latency, and most chips have hundreds of block RAMs on them (so each design module can use a few). However, it's very small - storing a single 640x480 colour image will use more block RAMs than many of the smaller chips have available.

Q13. When to use BRAM and DRAM?

Ans. Block Ram is a dedicated Ram that does not consume any additional LUT in your design whereas distributed Ram is built up with LUT. In terms of speed, the distributed RAM is faster than Block Rams. If not much RAM is needed, you can consider implementing it as a distributed Ram. Some synthesisers may even use distributed Ram if you specifically chose Block Ram but use a very small amount of a single block. As soon as you have large Ram blocks, it doesn't make much sense to use distributed Ram.

Q14. What are GTX transceivers on FPGA and why they are used?

Ans. A Gigabit Transceiver Xilinx (GTX) is a SerDes capable of operating at serial bit rates above 1 Gigabit/second. GTXs are used increasingly for data communications because they can run over longer distances, use fewer wires, and thus have lower costs than parallel interfaces with equivalent data throughput. Like other SerDes, the primary function of the GTX is to transmit parallel data as a stream of serial bits and convert the serial bits it receives to parallel data. The most basic performance metric of a GTX is its serial bit rate, or line rate, which is the number of serial bits it can transmit or receive per second. Although there is no strict rule, GTXs can typically run at line rates of 1 Gigabit/second or more. GTXs have become the 'data highways' for data processing systems that demand a high in/out raw data input and output (e.g. video processing applications). They are becoming very common on FPGA - such programmable logic devices being especially well fitted for parallel data processing algorithms.

Q15. How is the Zynq FPGA series different from other FPGA 7 series?

Ans. Firstly, Zynq is an FPGA plus hard processor, making it an SoC. 28nm Zynq SoCs has the 28 nm/7 series FPGA fabric in them whereas the Zynq MPSoC has Ultrascale FPGA fabric. The main difference between the Zynq processing system and Microblaze based is that in Zynq systems, the processor section boots before FPGA fabric also these are hard cores. Zynq SoC can implement a Microblaze subsystem in its FPGA fabric. Of course, these are very high-level differences only. There's a bunch of other differences which can be found on Xilinx documentation. **UG535** is a bible for understanding Zynq architecture and all its components.

Q16. What are HR and HP banks in FPGA?

Ans. HR means "High Range" bank and HP means "High Performance" bank. HR and HP define the different voltage values I/O banks. Voltage values under High Range are 1.2v, 1.35v, 1.5v, 1.8v. Voltage values under High Performance are 1.2v, 1.35v, 1.5v, 1.8v, 2.5v, 3.3v.

Q17. How many ways are there to implement multiplier in FPGA?

Ans. There are five ways to implement multiplier on FPGA:

1. Combinational circuit
2. Sequential circuit (FSM)
3. Speciality Circuits (Booth's Algorithm)
4. Memory
5. Hard multiplier blocks.

Among the five, the hard multiplier blocks are the best way, memory method is preferred for smaller input widths.

Q18. What is a sequence detector?

Ans. A sequence detector accepts as input a string of bits either 0 or 1. Its output goes to 1 when a target sequence has been detected. There are two basic types **overlap and non-overlap**. In a sequence detector that allows overlap, the final bits of one sequence can be the start of another sequence. One example will be a 11011-sequence detector. It raises an output of 1 when the last five binary digits are 11011. At the point, a detector with overlap will allow the last two 1 bits to serve at the first of the next sequence. The sequence detector with no overlap allowed resets itself to the state when the sequence has been detected. Write the input sequence as 11011011011. After the initial sequence 11011 has been detected, the detector with no overlap resets and starts searching for the initial of the next sequence. The detector with overlap allowed begins with the final 11 of the previous sequences as ready to apply as the first 11 of the next sequences; the next bit it is looking for is the 0.

Q19. Explain the sequence detector 11011?

Ans. Sequence detector has two types –

1. Overlapping

2. Non-Overlapping

In overlapping some of the last bits can also be used for the start of detection of next sequence within the given bits.

For example, Let the sequence be 11011 and given bits 1101101101101101

Now let's work on an overlapping concept.

We have 5 bits here in 11011 hence we will have five states. Let them be A/B/C/D and E.

Initially, the state will be A.

Now

1. The incoming bit is 1 (from 1101101101101101) and it matches with the first bit of sequence hence jump to next B. Requirement(1011)

2. The incoming bit is 1 (from 1101101101101101)and it matches with the first bit of requirement hence jump to state C. Requirement (011)

3. The incoming bit is 0 (from 1101101101101101) and it matches with the first bit of requirement hence jump to state D. Requirement (11)

4. The incoming bit is 1 (from 1101101101101101) and it matches with the first bit of requirement hence jump to state E. Requirement (1)

5. The incoming bit is 0 (from 1101101101101101) and it matches with the first bit of requirement hence jump to state A. Requirement (). The output is 1 as we have found a sequence

To be clearer here is a table-

State	Has	Awaiting
A	--	11011
B	1	1011
C	11	011
D	110	11
E	1101	1

Notice that state C has 11 and requires 011. Now if it receives 1 instead of 0 then instead of resetting and going back to A, it will remain at C because C has 11 which can be used for starting of 11011. This is called Overlapping. Similarly, after state E we have restart to detect sequence then instead of starting again from A we will jump to C since it already has some bits which can serve as a starting point. Remember always jump to that state which can provide **maximum** starting bits of sequence. Here B has 1 which can also serve, but it isn't maximum.

Ok again for better understanding we have 11011 then

1101<u>1</u> can serve as starting bit, i.e. state B

110<u>11</u> can serve as starting bits, i.e. state C

11<u>011</u> cannot serve as starting bits since sequence doesn't start with 011

1<u>1011</u> cannot serve as starting bits since sequence doesn't start with 1011

Here is the state diagram for this sequence. I am pretty sure you must have understood Overlapping till now. If No! You can contact, or this state diagram should suffice.

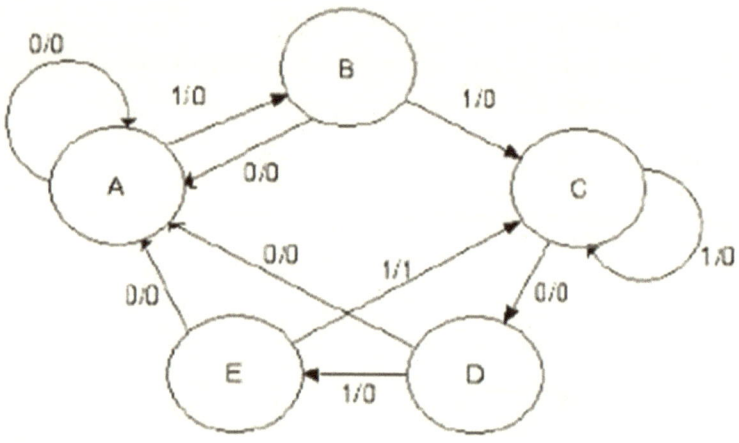

Figure 2.6: 11011 Mealy FSM

Let's go step by step

A. The idle state is A waiting for 1 which if it gets will jump to B else will remain to A

B. If receives 1 will jump to C else will jump back to idle A

C. If receives 1 will remain at itself as it has 11 to start with however if it receives 0 then it will jump to state D

D. If receives 0 will jump to state A else will jump to state E

E. If receives 0 will jump to A else will jump to C and output will be 1, which means that a sequence has been detected.

Now, how many FFs do we require to make this machine? We have 5bits here so by using this equation, we can find

$2^{x-1}<5<2^x$

Thus, we get X = 3 hence 3 FFs

To design in verilog here is the code for both Overlap and Non-Overlap.

//***//

module sequence_detector(sequence,overlap,detect,Clock,q);

input [15:0]sequence;

reg [15:0]temp;

reg bitin;

input overlap;

input [4:0]detect;

input Clock;

output reg q;

integer i=0;

reg [2:0]pstate;

parameter A = 3'b000, B = 3'b001, C = 3'b011, D = 3'b100, E = 3'b101;

initial begin

pstate<=A;

$monitor("Pstate=%d bit=%b q=%b",pstate,bitin,q);

end

always @ (posedgeClock)begin

if(i==0)

temp = sequence;

```verilog
i = i + 1;
end
always @ (posedgeClock)begin
bitin = temp[15];
temp=temp<<1;
if(overlap==1'b1)begin
case(pstate)
A:
if(bitin==1'b1)begin
pstate = B;
q = 0;
end
else begin
pstate = A;
q = 0;
end
B:
if(bitin==1'b1)begin
pstate = C;
q = 0;
end
else begin
pstate = A;
q = 0;
end
```

```
C:
if(bitin==1'b0)begin
pstate = D;
q = 0;
end
else begin
pstate = C;
q = 0;
end
D:
if(bitin==1'b1)begin
pstate = E;
q = 0;
end
else begin
pstate = A;
q = 0;
end
E:
if(bitin==1'b1)begin
pstate = C;
q = 1;
end
else begin
pstate = A;
```

```
        q = 0;
    end
    endcase
end
else if(overlap==1'b0) begin
    case(pstate)
    A:
        if(bitin==1'b1)begin
            pstate = B;
            q = 0;
        end
        else begin
            pstate = A;
            q = 0;
        end
    B:
        if(bitin==1'b1)begin
            pstate = C;
            q = 0;
        end
        else begin
            pstate = A;
            q = 0;
        end
    C:
```

```
if(bitin==1'b0)begin
pstate = D;
q = 0;
end
else begin
pstate = A;
q = 0;
end
D:
if(bitin==1'b1)begin
pstate = E;
q = 0;
end
else begin
pstate = A;
q = 0;
end
E:
if(bitin==1'b1)begin
pstate = A;
q = 1;
end
else begin
pstate = A;
q = 0;
```

```
            end
        endcase
    end
end
endmodule
```

and here is the **Test Bench**

```
//**********************************************************************//
module sequence();
reg Clock,overlap;
reg [4:0]detect;
reg [15:0]sequence;
wire q;

initial begin
Clock = 0;
overlap = 1; //1 for overlap 0 for non overlap
sequence = 16'b1101101101101101;
detect = 5'b11011;
end
always #2 Clock=!Clock;
sequence_detector yeah(sequence,overlap,detect,Clock,q);
endmodule
```

Q20. Why negative slack violation occurs?

Ans. If data path delay (arrival time) is greater than the destination clock path(required time), then the negative slack violation occurs.

Q21. What is the difference between synthesis and implementation timing report?

Ans. In a synthesis report, we get estimated time only. No actual routing delays are considered. Compare to delays from the synthesis report in the implementation report, and net delays are actual delays.

Q22. What is a racing condition?

Ans. A race condition is a flaw in a system or process that is characterised by an output that exhibits an unexpected dependence on the relative timing or order of events. The term originates with the idea of two signals racing each other attempting to influence the output first. When simulating a design with Verilog (or any other event-driven simulation language), engineers are often faced with two fundamentally different types of race conditions, hardware races and simulation induced races.

Q23. What are hardware races?

Ans. Hardware races typically occur in combinational logic due to the physical nature of their electronics.

For example, when the inputs to a logic gate change state, a finite delay transpires before the change is reflected in the gate's output.

Q24. What are simulation induced races?

Ans. Stimulation-induced races are not intrinsic to the design or its physics, but are natural, although undesirable, the consequence of the event-driven simulation algorithm used by Verilog. Because, the simulator processes events one at a time, it unavoidably serialises the events that occur in the same time slot. Hence, the design activity that in the actual hardware takes place concurrently is instead as a set of ordered actions by the simulator. This modelling deviation from the actual hardware induces additional races that are not present in the design but are an artefact of the simulator. These races can cause the simulator to simulate a faulty design when, in fact, the design is correct, or more dangerously, simulate a seemingly correct design when, in fact, the design is flawed. Frequently, this latter type of a race occurs because the designer's code relies, often unwittingly, on the specific ordering of the simulation algorithm. It is precisely for this reason that Verilog specifies that an event region must be processed in an arbitrary order, but every implementation will exhibit a certain order.

Q25. What is the extension of constraint file?

Ans. The extension of constraint file is.xdc (Xilinx Design Constraint) in Vivado software, and for ISE Design suite the extension of constraint file is.ucf(user constraint file).

Q26. What is the importance of the constraint file?

Ans. There are three classes of configuration information that need to be defined as constraints. These are:

1. **Specific to the Device and Board** - The pins used on a given device, on a given board.

2. **Specific to a Certain Device** - Including specifications for how to use internal features of the device.

3. **Specific to a Project or Design** - Includes requirements such as the allocation of a clock resource, and at speed, it must run. All three are implemented as constraints within constraint files. Constraint files can contain any number of different constraints, for any of the classes mentioned above. To ensure the portability of FPGA design, the most logical approach is to break the constraints into these three classes and store them in separate constraint files. Overall, we can say that by constraint file you tie your HDL code

I/O with FPGA's evaluation board I/O. Always refer schematic to find the available I/O's and their pin position.

Q27. What are TNS, WNS, THS, and WHS?

Ans. WNS = Worst Negative Slack

TNS = Total Negative Slack = sum of the negative slack paths

WHS = Worst Hold Slack

THS = Total Hold Slack = sum of the negative hold slack paths

Q28. How to deal with the warning of THS and WHS and with negative values of WNS, TNS, WHS, THS?

Ans. Unfortunately, the ways to resolve this are not "nice" but here are some:

1. Lower the clock speed. If you're at 100MHz now, dropping it to 80MHz or 70MHz will probably work.

2. If the design is heavily congested, reducing congestion (by reducing the amount of hardware) will help.

3. Rewriting the relevant bits of code with less work done in each cycle (ideal option, but potentially very time consuming).

And Before I suggest specific to any violation, there are many issues you need to overcome, and I hope after solving those most of the violations will get resolved.

1. Check_timing report section at the start of the file you have shared indicates that there are many sequential cells without the clock. Identify those and provide proper constraints.

2. There are many registers with multiple clock definitions on it, rectify those issues.

3. Add INPUT and OUTPUT delay constraints for all top-level ports.

4. There are CDC paths under violations, check for those. Use proper CDC circuit for proper data transfer. Write proper constraints if those are valid timing paths.

5. Also, check if there are any asynchronous clocks and as peruse case you don't want analysis. If so, define proper clock groups for the same.

Note: In Vivado all clocks consider as Synchronous unless specified by the user as asynchronous.

You can use report clock interaction and check if any path comes under the unsafe or partially timed section.

Overall this is under constraint design; once you constrained this design properly, most of the violations will get resolved by the tool itself. After proper baselining (Constraining) if you still face any issue specific to clock domain or CDC. Then you can go ahead and try a different strategy, floorplanning or RTL changes to overcome the violations.

Q29. What are different I/O standard available for FPGA?

Ans.
1. LVCMOS25
2. LVCMOS33
3. LVCMOS12
4. HSTL_I (High-Speed Terminated Logic)
5. HSTL_I_18
6. HSTL_I_12
7. HSTL_I_DCI_18
8. HSTL_II
9. HSTL_II_18
10. HSTL_II_DCI
11. HSTL_II_DCI_18
12. SSTL135 (Stub Series Terminated Logic)
13. SSTL135_DCI

14. SSTL15

15. SSTL15_DCI

16. SSTL18_I

17. SSTL18_I_DCI

18. SSTL18_II

19. SSTL18_II_DCI

20. HSUL_12 (High-Speed Unterminated Logic)

21. HSUL_12_DCI

22. LVCMOS18

23. LVDS (Low Voltage Differential Signaling)

24. LVDS_25

25. RSDS (Reduced Swing Differential Signaling)

26. Mini_LVDS_25

27. PPDS (Point to Point Differential Signaling)

28. TMDS (Transition Minimized Differential Signaling)

29. BLVDS (Bus LVDS)

Q30. Difference between DDR ODT and DCI?

Ans. **DCI (Digitally controlled impedance)** is used to make sure the FPGA driver and termination impedances are always at their expected value during operation and accounts for any process variation in the device as well as temperature and voltage changes. This is derived from internal logic and the 240-ohm resistor connected on the VRP pin.

ODT (On Die Termination) is the actual termination value at the receiver on the DDR interface. The DDR device derives this from a mode register setting and the 240-Ohm ZQ resistor. Also, worth noting that DDR devices do have ZQ calibration to make sure it adjusts for temperature and voltage changes. The FPGA derives this from the IP settings and is constantly adjusted to remain accurate with the DCI feature. Wrapping this together, when the FPGA needs to have 40-ohm ODT enabled during a read, DCI makes sure the resistance present at the FPGA receivers is 40-ohms.

Q31. Difference between FPGA and ASIC and which one to use?

Ans. **FPGA:**

1. Reconfigurable circuit. FPGAs can be reconfigured with a different design. They even can reconfigure a part of the chip while remaining areas of a chip are still working! This feature is widely used in accelerated computing in data centres.

2. Design is specified generally using hardware description languages (HDL) such as VHDL or Verilog.

3. Easier entry-barrier. One can get started with FPGA development for as low as USD 30.

4. Not suited for very high-volume mass production.

5. Less energy-efficient requires more power for the same function which ASIC can achieve at lower power.

6. Limited in operating frequency compared to ASIC of similar process node. The routing and configurable logic eat up timing margin in FPGAs.

7. Analog designs are not possible with FPGAs. Although FPGAs may contain specific analog hardware such as PLLs, ADC etc. they are not much flexible to create, for example, RF transceivers.

8. FPGAs are highly suited for applications such as Radars, Cell Phone Base Stations etc. where the current design might need to be upgraded to use a better algorithm or to a better design. In these applications, the high cost of FPGAs is not the deciding factor. Instead, programmability is the deciding factor.

9. Preferred for prototyping and validating a design or concept. Many ASICs are prototyped using FPGAs themselves! Major processor manufacturers themselves use FPGAs to validate their System-on-Chips (SoCs). It is easier to make sure design is working as intended using FPGA prototyping.

10. FPGA designers generally do not need to care for back-end design. Everything is handled by synthesis and routing tools which make sure the design works as described in the RTL code and meets timing. So, designers can focus on getting the RTL design done.

ASIC:

1. Permanent circuitry. Once the application-specific circuit is taped-out into silicon, it cannot be changed. The circuit will work the same for its complete operating life.

2. Same as for FPGA. Design is specified using HDL such as Verilog, VHDL etc.

3. Very high entry-barrier in terms of cost, learning curve, liaising with semiconductor foundry etc. Starting ASIC development from scratch can cost well into millions of dollars.

4. Suited for very high-volume mass production.

5. Much more power-efficient than FPGAs. Power consumption of ASICs can be very minutely controlled and optimized.

6. ASIC fabricated using the same process node can run at a much higher frequency than FPGAs since its circuit is optimized for its specific function.

7. ASICs can have complete analog circuitry, for example, Wi-Fi transceiver, on the same die along with microprocessor cores. This is the advantage which FPGAs lack.

8. ASICs are not suited for application areas where the design might need to be upgraded frequently or once-in-a-while.

9. It is not recommended to prototype a design using ASICs unless it has been validated. Once the silicon has been taped out, almost nothing can be done to fix a design bug (exceptions apply).

10. ASIC designers need to care for everything from RTL down to reset tree, clock tree, physical layout and routing, process node, manufacturing constraints (DFM), testing constraints (DFT) etc. Generally, each of the mentioned areas is handled by a different specialist person.

Q32. What is single-ended signalling?

Ans. Single-ended signalling is the simplest and most commonly used method of transmitting electrical signals over wires. One wire carries a varying voltage that represents the signal, while the other wire is connected to a reference voltage, usually ground. The main alternative to single-ended signalling is called differential signalling. There is also the historical alternative of ground return, rarely used today. Single-ended signalling is less expensive to implement than differential, but it cannot reject noise caused by:

1. Differences in ground voltage level between transmitting and receiving circuits

2. Induction picked up on the signal wire the main advantage of single-ended over differential signalling is that fewer wires are needed to transmit multiple signals. If there are n signals, then there are n+1 wire - one for each signal and one for ground. (Differential signalling uses at least 2n wires.)

A disadvantage of single-ended signalling is that the return currents for all the signals use the same conductor (even if separate ground wires are used, the grounds are inevitably connected at each end), and this can sometimes cause interference ("crosstalk") between the signals.

Single-ended signalling is widely used and can be seen in numerous common transmission standards, including RS232, PS/2, I2C, TTL, CMOS, ECL etc.

Q33. What is differential signalling?

Ans. Differential signalling is a method for electrically transmitting information using two complementary signals. The technique sends the same electrical signal as a differential pair of signals, each in its conductor. The pair of conductors can be wires (typically twisted together) or traces on a circuit board. The receiving circuit responds to the electrical difference between the two signals, rather than the difference between a single wire and ground. The opposite technique is called single-ended signalling. Differential pairs are usually found on printed circuit boards, in twisted-pair and ribbon cables, and connectors. Provided that the source and receiver impedances in the differential signalling circuit are equal, external electromagnetic interference tends to affect both conductors identically. Since the receiving circuit only detects the difference between the wires, the technique resists electromagnetic noise compared to one conductor with an unpaired reference (ground). The technique work for both analog signalling, as in balanced audio—and digital signalling, as in RS-422, RS-485, Ethernet over twisted pair, PCI Express, DisplayPort, HDMI, and USB.

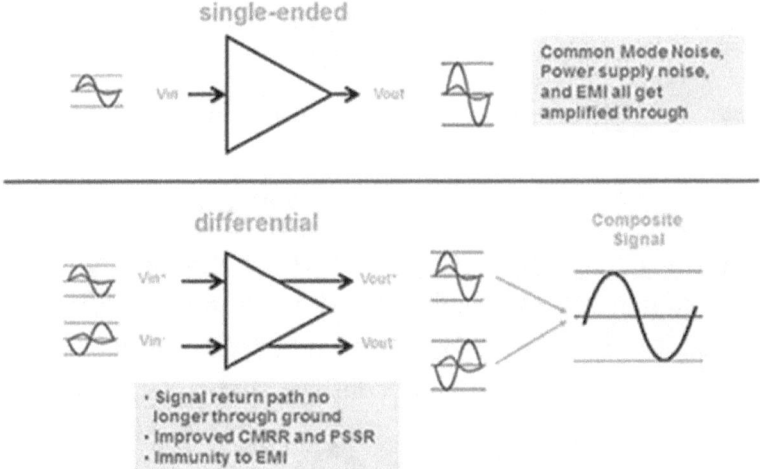

Figure 2.7: Single-ended and differential signalling

Q34. What are the different series of Xilinx FPGA?

Ans. 45nm: Spartan6

28nm: Artix7, Virtex7, Kintex7, Spartan7

20nm: KintexUltrascale, VirtexUltrascale.

16nm: KintexUltrascale+, VirtexUltrascale+

Devices	ARTIX7	KINTEX7	VIRTEX7
USP	Lowest Power & Cost	Industry's Best Price/ Performance	Industry's Highest System Performance
Logic Cells	20K – 355K	30K – 410K	285K – 2000K
DSP Slices	40 – 700	120 – 1540	700 – 3960
Max. Transceivers	4	16	80
Transceiver Performance	3.75Gbps	6.6Gbps 10.3Gbps	10.3Gbps 13.1Gbps 28Gbps
Memory Performance	800Mbps	2133Mbps	2133Mbps
Max. SelectIO	450	500	1200
SelectIO Voltages	3.3V & above	3.3V and below 1.8V and below	3.3V and below 1.8V and below

Q35. What is SDK (Software Development Kit) in Vivado software?

Ans. The Xilinx® Software Development Kit (SDK) provides an environment for creating software platforms and applications targeted for Xilinx embedded processors. SDK works with hardware designs created with Vivado®. ... Application build configuration and automatic Makefile generation.

Steps to launch SDK from Vivado:

1. Click on the File tab.

2. In the File tab, click on the Export option.

3. In Export option, select the Export Bitstream File

After exporting the Bitstream file, again go to the File tab and click and launch SDK.

Q36. What is FIFO and Why is it used?

Ans. The acronym FIFO stands for First In First Out. FIFOs are used everywhere in FPGA and ASIC designs; they are one of the basic building blocks, and they are very handy! FIFOs can be used for any of these purposes:

1. Crossing clock domains

2. Buffering data before sending it off-chip (e.g. to DRAM or SRAM).

3. Buffering data for the software to look at a later time.

4. Storing data for later processing.

A FIFO can be thought of a one-way tunnel that cars can drive through. At the end of the tunnel is a toll with a gate. Once the gate opens, the car can leave the tunnel. If that gate never opens and more cars keep entering the tunnel, eventually the tunnel will fill up with cars. This is called FIFO Overflow, and in general, it's not a good thing. How deep the FIFO is can be thought of as the length of the tunnel. The deeper the FIFO, the more data can fit into it before it overflows. FIFOs also have a width, which represents the width of the data (in the number of bits) that enters the FIFO.

The two rules of FIFOs:

1. Never write to a full FIFO (overflow)

2. Never read from an empty FIFO (underflow)

FIFOs themselves can be made up of dedicated pieces of logic inside your FPGA or ASIC, or they can be created from Flip-Flops (distributed registers). Which one of these two synthesis tools will use is entirely dependent on the FPGA vendor that you are using and how you structure your code? Just know that when you use the dedicated pieces of logic, they have better performance than having a register-based FIFO.

Q37. What are synchronous and asynchronous FIFO?

Ans. A synchronous FIFO would use the same clocks for reading and writing.

Asynchronous uses different clocks.

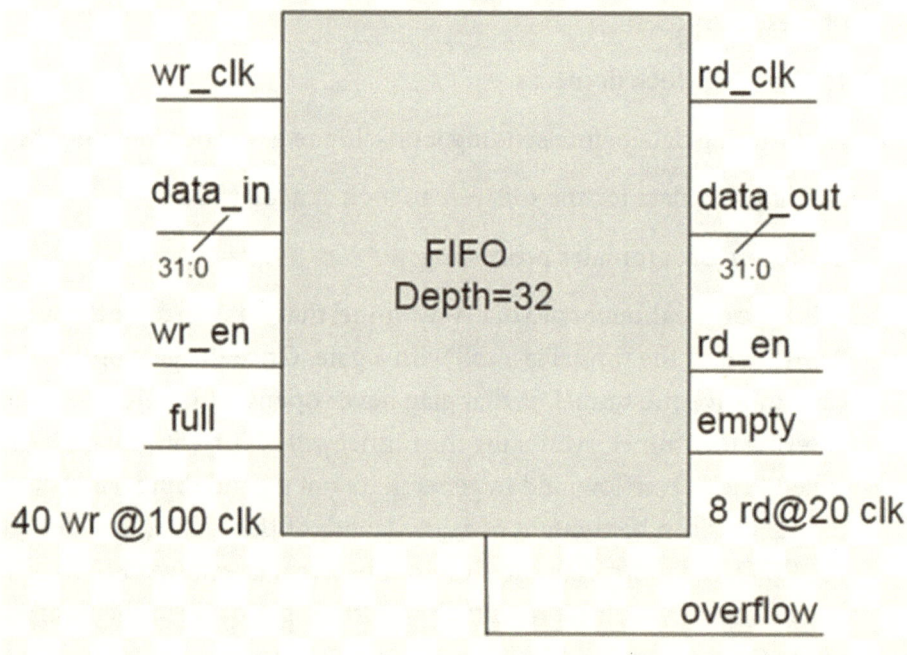

Figure 2.8: Asynchronous FIFO

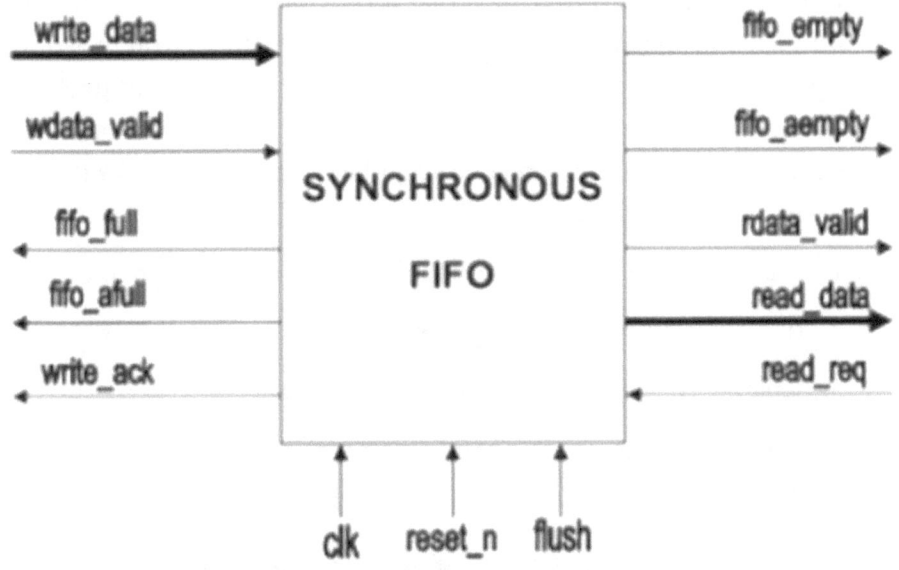

Figure 2.9: Synchronous FIFO

Q38. What are the different types of interconnect used in FPGA architecture?

Ans. **1. Anti-Fuse interconnect:** These are highly reliable but one time programmable and expensive.

2. Flash interconnect: these are highly reliable, reprogrammable but more expensive than SRAM FPGA.

3. SRAM interconnect: These are not highly reliable, reprogrammable and highest density, lowest cost. Xilinx provides Flash Types CPLD and FPGA, SRAM FPGA. Altera provides Flash Types CPLD and SRAM FPGA.

Q39. What is the AXI interface?

Ans. Advanced eXtensible Interface, or AXI, is part of ARM's AMBA specifications. The AXI is a point to point interconnect that designed for high performance, high-speed microcontroller systems. The AXI protocol is based on a point to point interconnect to avoid bus sharing and therefore allow higher bandwidth and lower latency. AXI is arguably the most popular of all AMBA interface interconnect. The essence of the AXI protocol is that it provides a framework for how different blocks inside each chip communicate with each other. It offers a procedure before anything is transmitted so that the communication is clear and uninterrupted. That way, different components can talk to each other without stepping on each other. The procedure for the AXI protocol is as follows:

1. Master & slave must "handshake" to confirm valid signals.
2. Transmission of control signal must be in separate phases.
3. Separate channels for transmission of signals.
4. The continuous transfer can be accomplished through burst-type communication.

Q40. What are the different types of AXI Interface and mention their difference?

Ans. Three flavours: AXI4, AXI4-Lite, AXI4-Stream. All three share the same handshake rules and signal naming.

1. AXI4:
    a. **Dedicated for**: high-performance and memory-mapped systems.
    b. **Burst(data beta):** up to 256
    c. **Data width:** 32 to 1024 bits
    d. **Applications(examples):** Embedded, memory
2. AXI4-Lite:
    a. **Dedicated for**: register-style interfaces (area efficient implementation)
    b. **Burst(data beta):** 1 bit
    c. **Data width:** 32 or 64 bits
    d. **Applications(examples):** Small footprint control logic.

3. AXI4-Stream:

   a. **Dedicated for:** Non-address-based IP (PCIe, Filters, etc.)

   b. **Burst(data beta):** Unlimited

   c. **Data width:** Any number of bytes

   d. **Applications(examples):** DSP, video, communication

Q41. What is DCM's and why they are used?

Ans. Digital clock manager (DCM) is a fully digital control system that uses feedback to maintain clock signal characteristics with a high degree of precision despite normal variations in operating temperature and voltage. That is the clock output of DCM is stable over a wide range of temperature and voltage, and skew associated with DCM is minimal, and all phases of input clock can be obtained. The output of DCM coming from global buffer can handle more load.

Q42. Compare PLL and DLL?

Ans. **PLL:** PLLs have disadvantages that make their use in high-speed designs problematic, particularly when both high performance and high reliability are required. The PLL voltage-controlled oscillator (VCO) is the greatest source of problems. Variations in temperature, supply voltage, and manufacturing process affect the stability and operating performance of PLLs.

DLLs, however, are immune to these problems. A DLL in its simplest form inserts a variable delay line between the external clock and the internal clock. The clock tree distributes the clock to all registers and then back to the feedback pin of the DLL. The control circuit of the DLL adjusts the delays so that the rising edges of the feedback clock align with the input clock. Once the edges of the clocks are aligned, the DLL is locked, and both the input buffer delay and the clock skew are reduced to zero.

Advantages:

- Precision
- Stability
- Power management
- Noise sensitivity
- Jitter performance

PART III

Protocols

Q1. Explain the UART protocol?

Ans. In UART communication, two UARTs communicate directly with each other. The transmitting UART converts parallel data from a controlling device like a CPU into serial form, transmits it in serial to the receiving UART, which then converts the serial data back into parallel data for the receiving device. Only two wires are needed to transmit data between two UARTs. Data flows from the Tx pin of the transmitting UART to the Rx pin of the receiving UART.

UARTs transmit data asynchronously, which means there is no clock signal to synchronize the output of bits from the transmitting UART to the sampling of bits by the receiving UART. Instead of a clock signal, the transmitting UART adds start and stop bits to the data packet being transferred. These bits define the beginning and end of the data packet, so the receiving UART knows when to start reading the bits. When the receiving UART detects a start bit, it starts to read the incoming bits at a specific frequency known as the baud rate. Baud rate is a measure of the speed of data transfer, expressed in bits per second (bps).

Both UARTs must operate at about the same baud rate. The baud rate between the transmitting and receiving UARTs can only differ by about 10% before the timing of bits gets too far off. Both UARTs must also be configured to transmit and receive the same data packet structure.

Figure 3.1:

Wires Used	2
Maximum Speed	Any Speed up to 115200, usually 9600 baud
Synchronous or Asynchronous?	Asynchronous
Serial or Parallel?	Serial
Max no. of Masters	1
Max no. of Slaves	1

The UART that is going to transmit data receives the data from a data bus. The data bus is used to send data to the UART by another device like a CPU, memory, or microcontroller. Data is transferred from the data bus to the transmitting UART in parallel form. After the transmitting UART gets the parallel data from the data bus, it adds a start bit, a parity bit, and a stop bit, creating the data packet. Next, the data packet is output serially, bit by bit at the Tx pin. The receiving UART reads the data packet bit by bit at its Rx pin. The receiving UART then converts the data back into parallel form and removes the start bit, parity bit, and stop bits. Finally, the receiving UART transfers the data packet in parallel to the data bus on the receiving end:

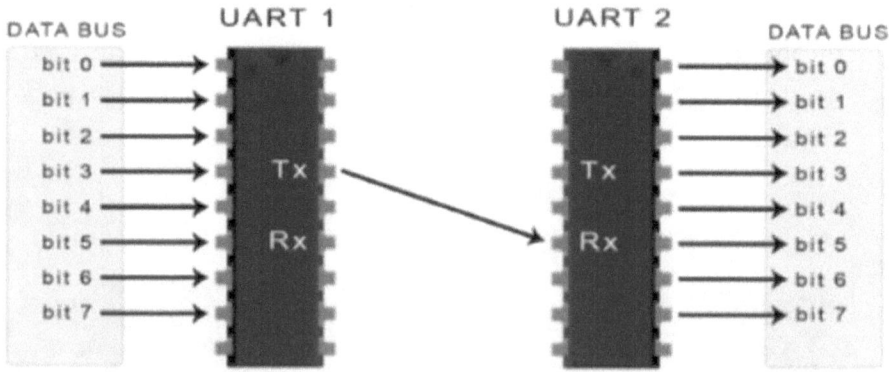

Figure 3.2:

UART transmitted data is organized into *packets*. Each packet contains 1 start bit, 5 to 9 data bits (depending on the UART), an optional *parity* bit, and 1 or 2 stop bits:

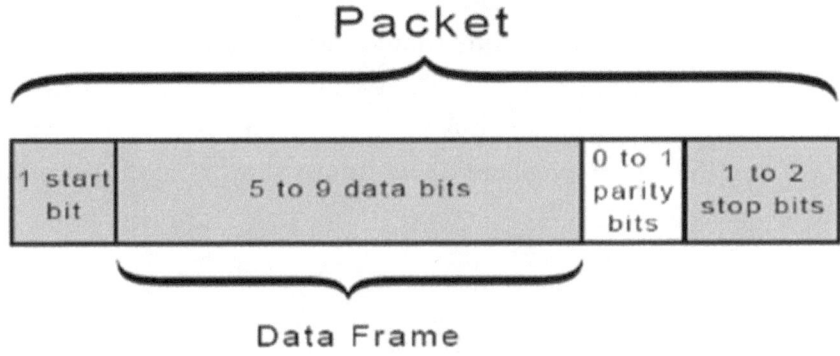

Figure 3.3:

START BIT

The UART data transmission line is normally held at a high voltage level when it's not transmitting data. To start the transfer of data, the transmitting UART pulls the transmission line from high to low for one clock cycle. When the receiving UART detects the high to low voltage transition, it begins reading the bits in the data frame at the frequency of the baud rate.

DATA FRAME

The data frame contains the actual data to be transferred. It can be 5 bits up to 8 bits long if the parity bit is used. If no parity bit is used, the data frame can be 9 bits long. In most cases, the data is sent with the least significant bit first.

PARITY

Parity describes the evenness or oddness of a number. The parity bit is a way for the receiving UART to tell if any data has changed during transmission. Bits can be changed by electromagnetic radiation, mismatched baud rates, or long-distance data transfers. After the receiving UART reads the data frame, it counts the number of bits with a value of 1 and checks if the total is an even or odd number. If the parity bit is a 0 (even parity), the 1 bit in the data frame should total to an even number. If the parity bit is a 1 (odd parity), the 1 bit in the data frame should total to an odd number. When the parity bit matches the data, the UART knows that the transmission was free of errors. But if the parity bit is a 0, and the total is odd, or the parity bit is a 1, and the total is even, the UART knows that bits in the data frame have changed.

STOP BITS

To signal the end of the data packet, the sending UART drives the data transmission line from a low voltage to a high voltage for at least two-bit durations.

STEPS OF UART TRANSMISSION

1. The transmitting UART receives data in parallel from the data bus:

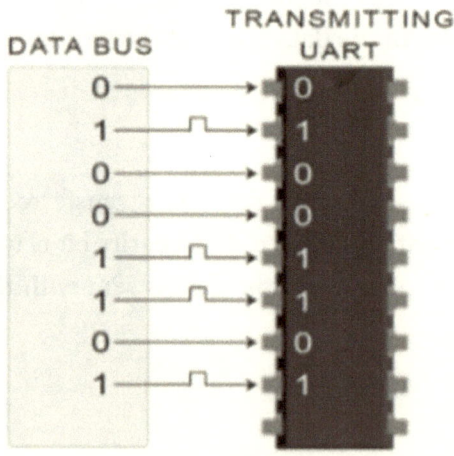

Figure 3.4:

2. The transmitting UART adds the start bit, parity bit, and the stop bit(s) to the data frame:

Figure 3.5:

3. The entire packet is sent serially from the transmitting UART to the receiving UART. The receiving UART samples the data line at the pre-configured baud rate:

Figure 3.6:

4. The receiving UART discards the start bit, parity bit, and stop bit from the data frame:

Figure 3.7:

5. The receiving UART converts the serial data back into parallel and transfers it to the data bus on the receiving end:

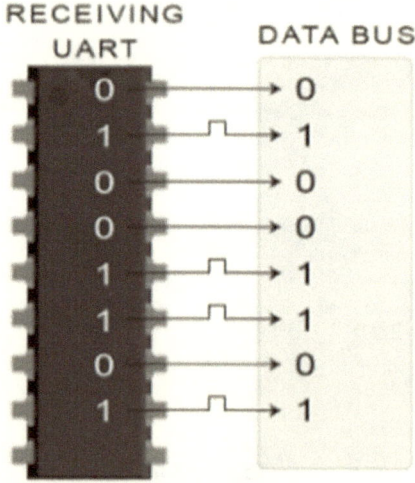

Figure 3.8:

No communication protocol is perfect, but UARTs are pretty good at what they do. Here are some pros and cons to help you decide whether they fit the needs of your project:

ADVANTAGES

1. Only uses two wires.
2. No clock signal is necessary.
3. Has a parity bit to allow for error checking.
4. The structure of the data packet can be changed if both sides are set up for it.
5. Well documented and widely used method.

DISADVANTAGES

1. The size of the data frame is limited to a maximum of 9 bits.
2. Doesn't support multiple slave or multiple master systems.
3. The baud rates of each UART must be within 10% of each other.

Q2. Explain how the I2C protocol works?

Ans. I2C combines the best features of SPI and UARTs. With I2C, you can connect multiple slaves to a single master (like SPI), and you can have multiple masters controlling single, or multiple slaves. This is useful when you want to have more than one microcontroller logging data to a single memory card or to display text to a single LCD. Like UART communication, I2C only uses two wires to transmit data between devices:

1. SDA (Serial Data) – The line for the master and slave to send and receive data.

2. SCL (Serial Clock) – The line that carries the clock signal. I2C is a serial communication protocol, so data is transferred bit by bit along a single wire (the SDA line).

Like SPI, I2C is synchronous, so the output of bits is synchronized to the sampling of bits by a clock signal shared between the master and the slave. The clock signal is always controlled by the master.

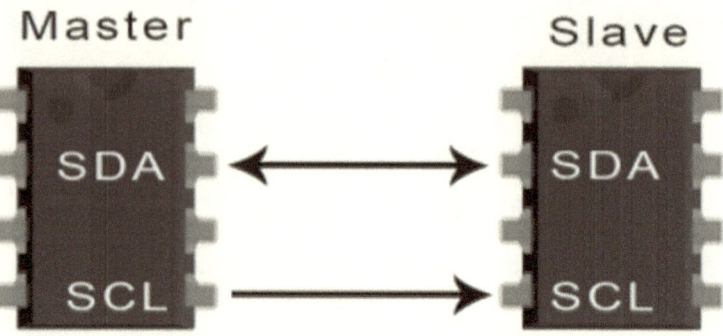

Figure 3.9:

Wires Used	2
Maximum Speed	Standard Mode = 100kbps
	Fast Mode = 400kbps
	High Speed Mode = 3.4Mbps
	Ultra-Fast Mode = 5Mbps
Synchronous or Asynchronous?	Synchronous
Serial or Parallel?	Serial
Max no. of Masters	Unlimited
Max no. of Slaves	1008

With I2C, data is transferred in *messages*. Messages are broken up into *frames* of data. Each message has an address frame that contains the binary address of the slave, and one or more data frames that contain the data being transmitted. The message also includes start and stop conditions, read/write bits, and ACK/NACK bits between each data frame:

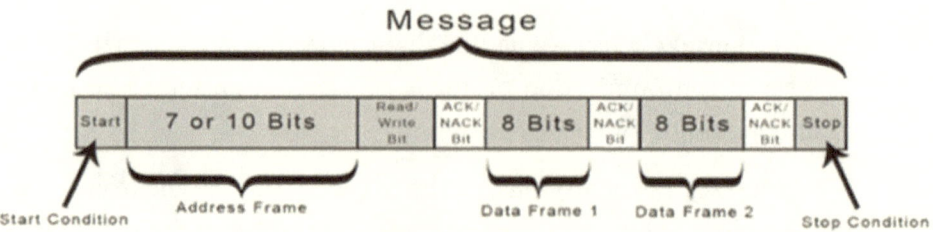

Figure 3.10:

Start Condition: The SDA line switches from a high voltage level to a low voltage level *before* the SCL line switches from high to low.

Stop Condition: The SDA line switches from a low voltage level to a high voltage level *after* the SCL line switches from low to high.

Address Frame: A 7-bit or 10-bit sequence unique to each slave that identifies the slave when the master wants to talk to it.

Read/Write Bit: A single bit specifying whether the master is sending data to the slave (low voltage level) or requesting data from it (high voltage level).

ACK/NACK Bit: Each frame in a message is followed by an acknowledge/no-acknowledge bit. If an address frame or data frame was successfully received, an ACK bit is returned to the sender from the receiving device.

ADDRESSING

I2C doesn't have slave select lines like SPI, so it needs another way to let the slave know that data is being sent to it, and not another slave. It does this by *addressing*. The address frame is always the first frame after the start bit in a new message.

The master sends the address of the slave it wants to communicate with to every slave connected to it. Each slave then compares the address sent from the master to its address. If the address matches, it sends a low voltage ACK bit back to the master. If the address doesn't match, the slave does nothing, and the SDA line remains high.

READ/WRITE BIT

The address frame includes a single bit at the end that informs the slave whether the master wants to write data to it or receive data from it. If the master wants to send data to the slave, the read/write bit is a low voltage level. If the master is requesting data from the slave, the bit is a high voltage level.

THE DATA FRAME

After the master detects the ACK bit from the slave, the first data frame is ready to be sent.

The data frame is always 8 bits long and sent with the most significant bit first. Each data frame is immediately followed by an ACK/NACK bit to verify that the frame has been received successfully. The ACK bit must be received by either the master or the slave (depending on who is sending the data) before the next data frame can be sent.

After all the data frames have been sent, the master can send a stop condition to the slave to halt the transmission. The stop condition is a voltage transition from low to high on the SDA line after a low to high transition on the SCL line, with the SCL line remaining high.

STEPS OF I2C DATA TRANSMISSION

1. The master sends the start condition to every connected slave by switching the SDA line from a high voltage level to a low voltage level *before* switching the SCL line from high to low:

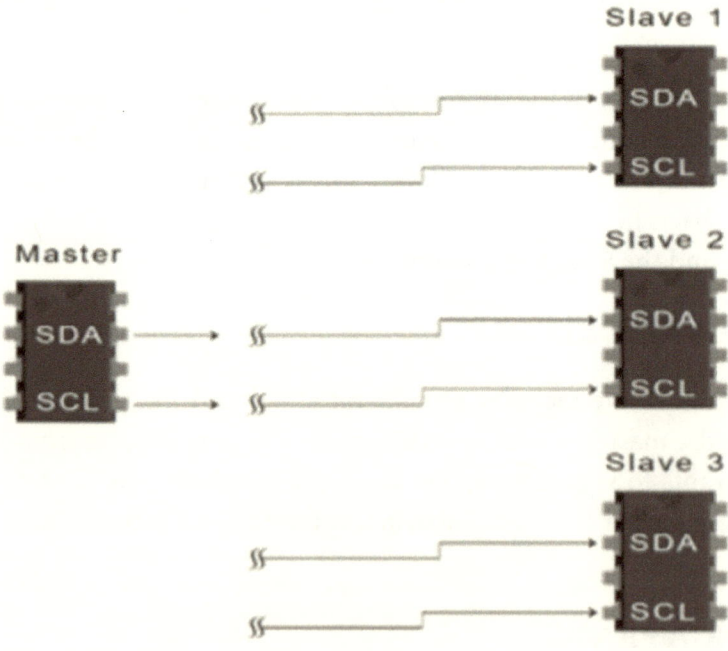

Figure 3.11:

2. The master sends each slave the 7 or 10-bit address of the slave it wants to communicate with, along with the read/write bit:

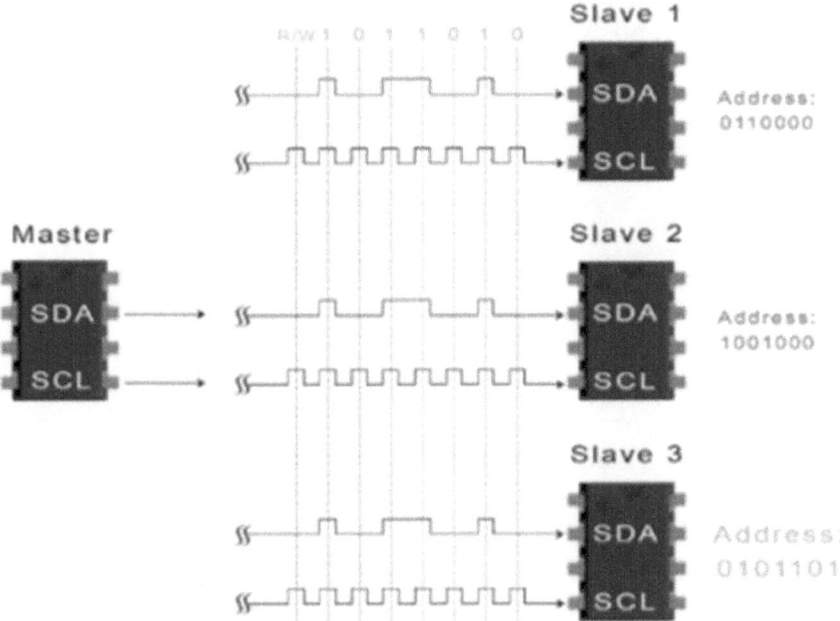

Figure 3.12:

3. Each slave compares the address sent from the master to its address. If the address matches, the slave returns an ACK bit by pulling the SDA line low for one bit. If the address from the master does not match the slave's address, the slave leaves the SDA line high.

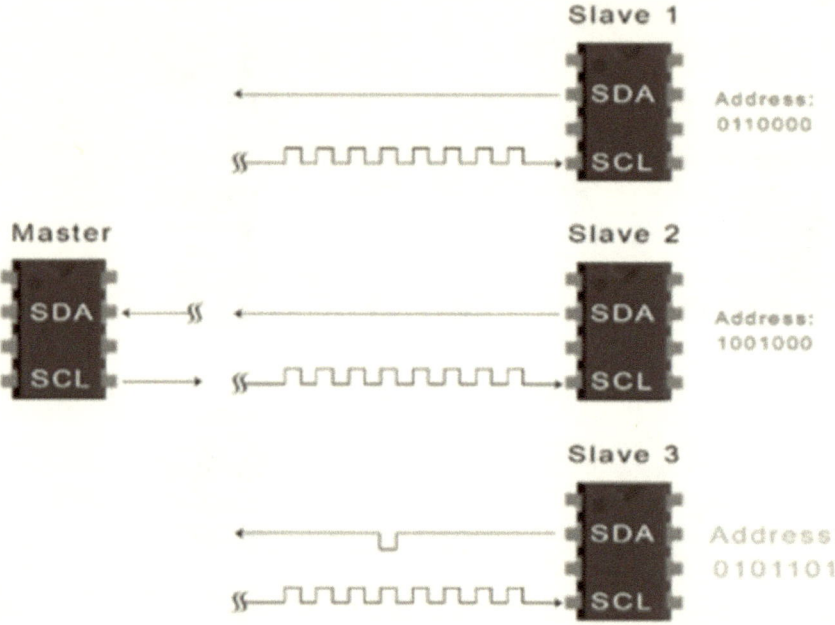

Figure 3.13:

4. The master sends or receives the data frame:

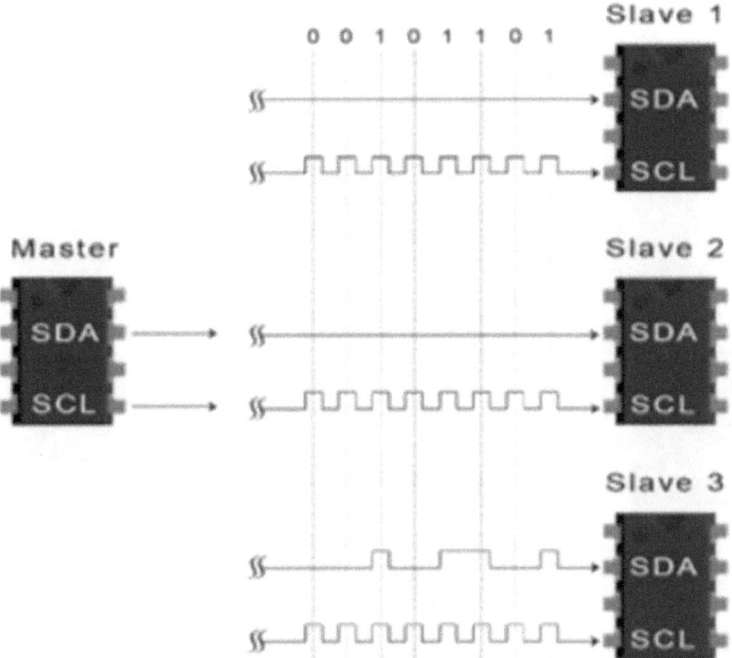

Figure 3.14:

5. After each data frame has been transferred, the receiving device returns another ACK bit to the sender to acknowledge successful receipt of the frame:

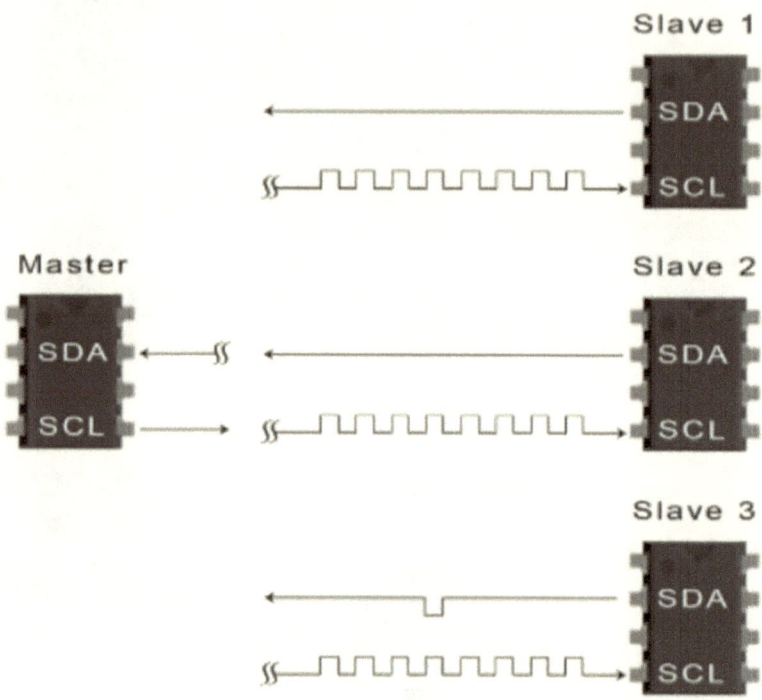

Figure 3.15:

6. To stop the data transmission, the master sends a stop condition to the slave by switching SCL high before switching SDA high:

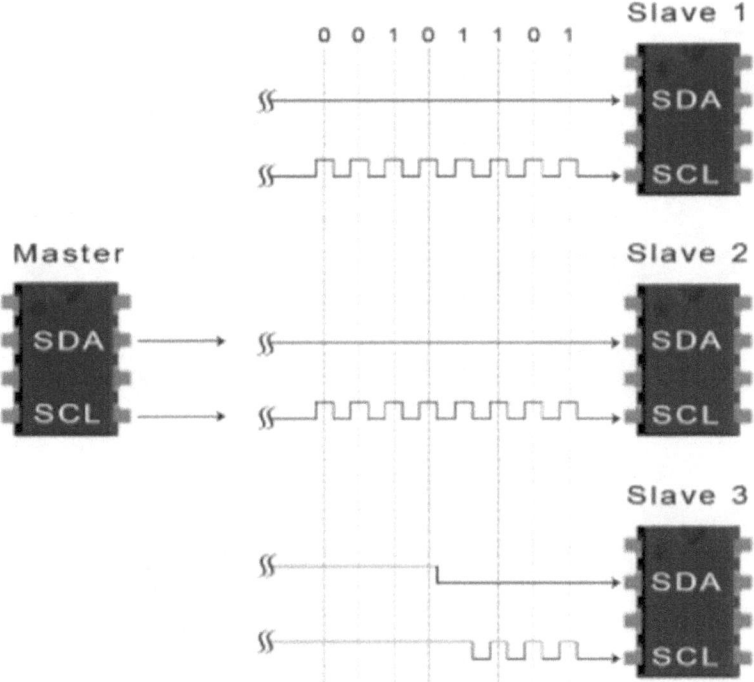

Figure 3.16:

SINGLE MASTER WITH MULTIPLE SLAVES

Because I2C uses addressing, multiple slaves can be controlled from a single master. With a 7-bit address, 128 (2^7) unique address are available. Using 10-bit addresses is uncommon but provides 1,024 (2^{10}) unique addresses. To connect multiple slaves to a single master, wire them like this, with 4.7K Ohm pull-up resistors connecting the SDA and SCL lines to Vcc:

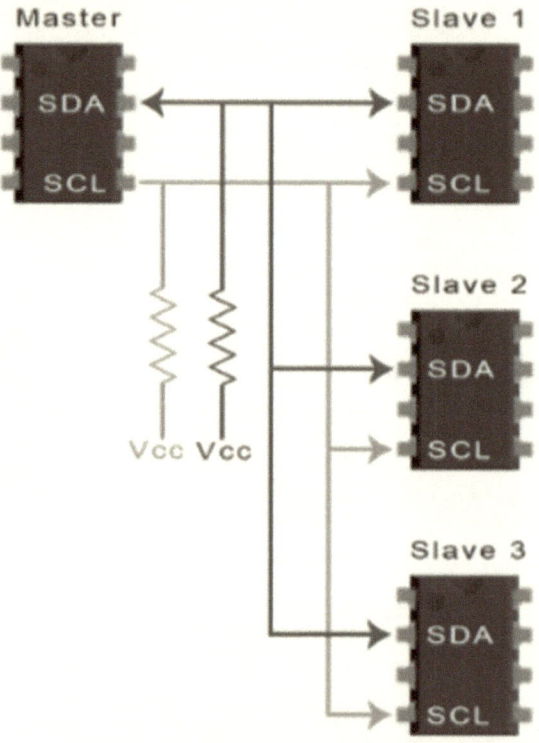

Figure 3.17:

MULTIPLE MASTERS WITH MULTIPLE SLAVES

Multiple masters can be connected to a single slave or multiple slaves. The problem with multiple masters in the same system comes when two masters try to send or receive data at the same time over the SDA line. To solve this problem, each master needs to detect if the SDA line is low or high before transmitting a message. If the SDA line is low, this means that another master has control of the bus, and the master should wait to send the message. If the SDA line is high, then it's safe to transmit the message. To connect multiple masters to multiple slaves, use the following diagram, with 4.7K Ohm pull-up resistors connecting the SDA and SCL lines to Vcc:

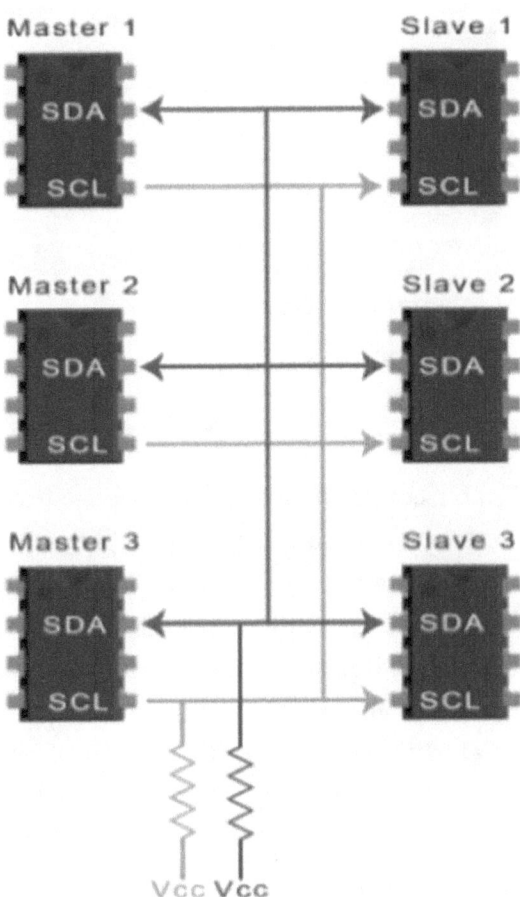

Figure 3.18:

ADVANTAGES AND DISADVANTAGES OF I2C

There is a lot to I2C that might make it sound complicated compared to other protocols, but there are some good reasons why you may or may not want to use I2C to connect to a device:

ADVANTAGES

1. Only uses two wires.
2. Supports multiple masters and multiple slaves.
3. ACK/NACK bit confirms that each frame is transferred successfully.
4. Hardware is less complicated than with UARTs.
5. Well known and widely used protocol.

DISADVANTAGES

1. Slower data transfer rate than SPI.
2. The size of the data frame is limited to 8 bits.
3. More complicated hardware needed to implement than SPI.

Q3. Explain how the SPI protocol works?

Ans. SPI is a common communication protocol used by many different devices. For example, SD card modules, RFID card reader modules, and 2.4 GHz wireless transmitter/receivers all use SPI to communicate with microcontrollers. One unique benefit of SPI is the fact that data can be transferred without interruption. Any number of bits can be sent or received in a continuous stream. With I2C and UART, data is sent in packets, limited to a specific number of bits. Start and stop conditions define the beginning and end of each packet, so the data is interrupted during transmission. Devices communicating via SPI are in a master-slave relationship. The master is the controlling device (usually a microcontroller), while the slave (usually a sensor, display, or memory chip) takes instruction from the master. The simplest configuration of SPI is a single master, single slave system, but one master can control more than one slave.

1. **MOSI (Master Output/Slave Input)** – Line for the master to send data to the slave.

2. **MISO (Master Input/Slave Output)** – Line for the slave to send data to the master.

3. **SCLOCK (Clock)** – Line for the clock signal.

4. **SS/CS (Slave Select/Chip Select)** – Line for the master to select which slave to send data to.

THE CLOCK

The clock signal synchronizes the output of data bits from the master to the sampling of bits by the slave. One bit of data is transferred in each clock cycle, so the speed of data transfer is determined by the frequency of the clock signal. SPI communication is always initiated by the master since the master configures and generates the clock signal.

Any communication protocol where devices share a clock signal is known as *synchronous*. SPI is a synchronous communication protocol. There are also *asynchronous* methods that don't use a clock signal. For example, in UART communication, both sides are set to a pre-configured baud rate that dictates the speed and timing of data transmission.

The clock signal in SPI can be modified using the properties of *clock polarity* and *clock phase*. These two properties work together to define when the bits are output and when they are sampled. Clock polarity can be set by the master to allow for bits to be output and sampled on either the rising or falling edge of the clock cycle. Clock phase can be set for output and sampling to occur on either the first edge or the second edge of the clock cycle, regardless of whether it is rising or falling.

SLAVE SELECT

The master can choose which slave it wants to talk to by setting the slave's CS/SS line to a low voltage level. In the idle, non-transmitting state, the slave select line is kept at a high voltage level. Multiple CS/SS pins may be available on the master, which allows for multiple slaves to be wired in parallel. If only one CS/SS pin is present, multiple slaves can be wired to the master by daisy-chaining.

MULTIPLE SLAVES

SPI can be set up to operate with a single master and a single slave, and it can be set up with multiple slaves controlled by a single master. There are two ways to connect multiple slaves to the master. If the master has multiple slave select pins, the slaves can be wired in parallel like this:

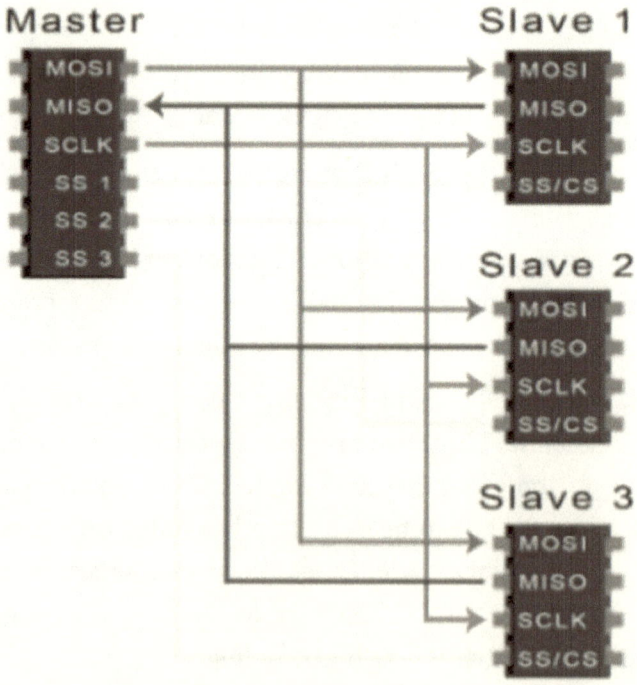

Figure 3.19:

If only one slave select pin is available, the slaves can be daisy-chained like this:

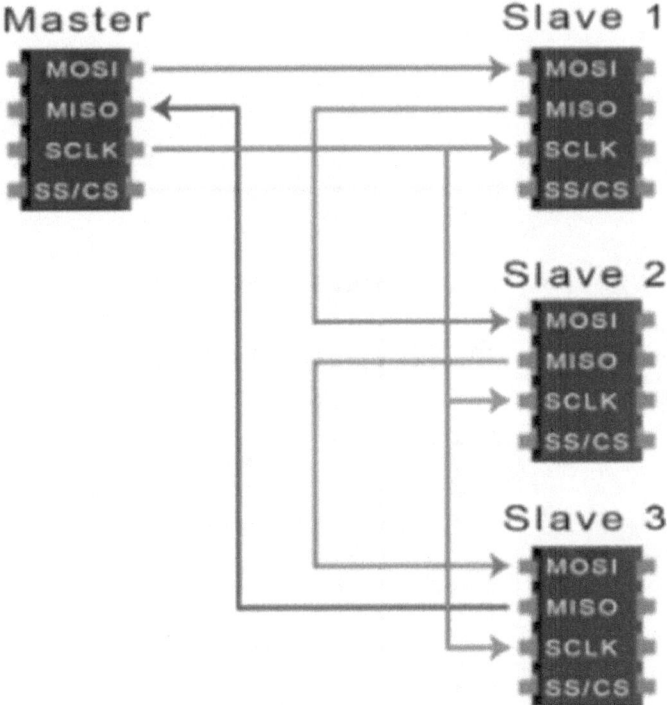

Figure 3.20:

MOSI AND MISO

The master sends data to the slave bit by bit, in serial through the MOSI line. The slave receives the data sent from the master at the MOSI pin. Data sent from the master to the slave is usually sent with the most significant bit first.

The slave can also send data back to the master through the MISO line in serial. The data sent from the slave back to the master is usually sent with the least significant bit first.

STEPS OF SPI DATA TRANSMISSION

1. The master outputs the clock signal:

Figure 3.21:

2. The master switches the SS/CS pin to a low voltage state, which activates the slave:

Figure 3.22:

3. The master sends the data one bit at a time to the slave along the MOSI line. The slave reads the bits as they are received:

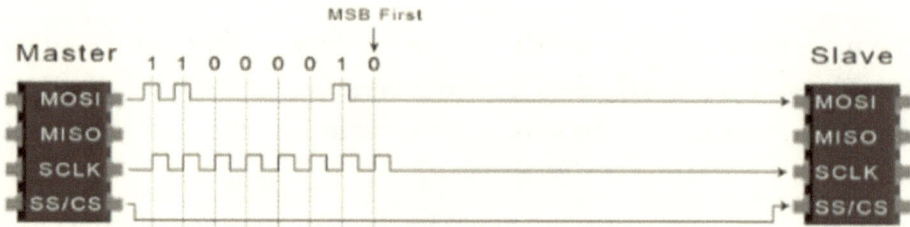

Figure 3.23:

4. If a response is needed, the slave returns data one bit at a time to the master along the MISO line. The master reads the bits as they are received:

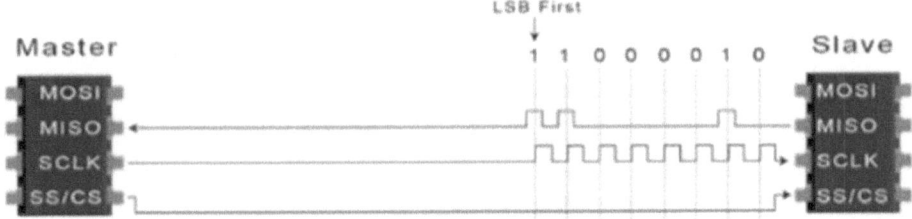

Figure 3.24:

ADVANTAGES AND DISADVANTAGES OF SPI

There are some advantages and disadvantages to using SPI, and if given a choice between different communication protocols, you should know when to use SPI according to the requirements of your project:

ADVANTAGES

1. No, start and stop bits so that the data can be streamed continuously without interruption.
2. No complicated slave addressing system like I2C.
3. Higher data transfer rate than I2C (almost twice as fast).
4. Separate MISO and MOSI lines so that data can be sent and received at the same time.

DISADVANTAGES

1. Uses four wires (I2C and UARTs use two).
2. No acknowledgement that the data has been successfully received (I2C has this).
3. No form of error checking like the parity bit in UART.
4. Only allows for a single master.

PART IV
STATIC TIMING ANALYSIS

Q1. What is Setup time and Hold Time? How can we remove setup and hold time violation?

Ans. Setup-time: The time interval before the active transition of the clock signal during which the data input(D, J or K) must be maintained.

Hold time: The time interval after the active transition of the clock signal during which the data input (D, J or K) must be maintained.

These timings for the data at the synchronous input (Let say D) must be stable before the active edge of the clock so that the data can be stored successfully in the storage device.

Setup violations can be fixed by either slowing down the clock (increase the period) or by decreasing the delay of the data path logic. The amount of time the data at the synchronous input (D) must be stable after the active edge of clock.so that the data can be stored successfully in the storage device.

Hold violations can be fixed by increasing the delay of the data path or by decreasing the clock uncertainty, which we also named as skew if specified in the design.

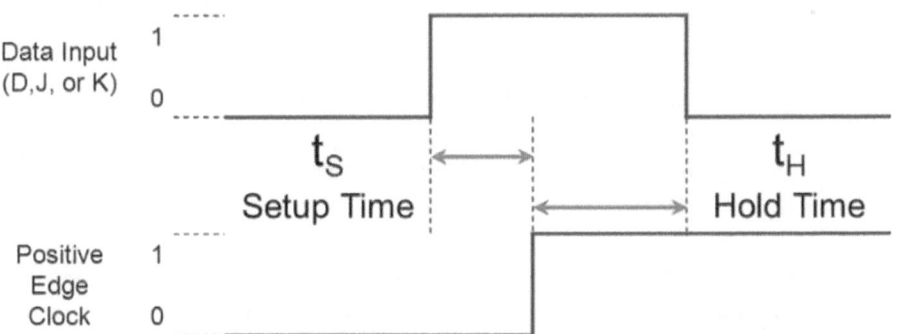

Figure 4.1: Set-up time and hold time

Q2. What is Metastability?

Ans. Whenever there are setup and hold time violations in any flip flop, it enters a state where its output is unpredictable, this state is known as a

metastable state or **quasi-stable state**, at the end of metastable state, the flip flop settles down to either 1 or 0. This whole process is known as **metastability**.

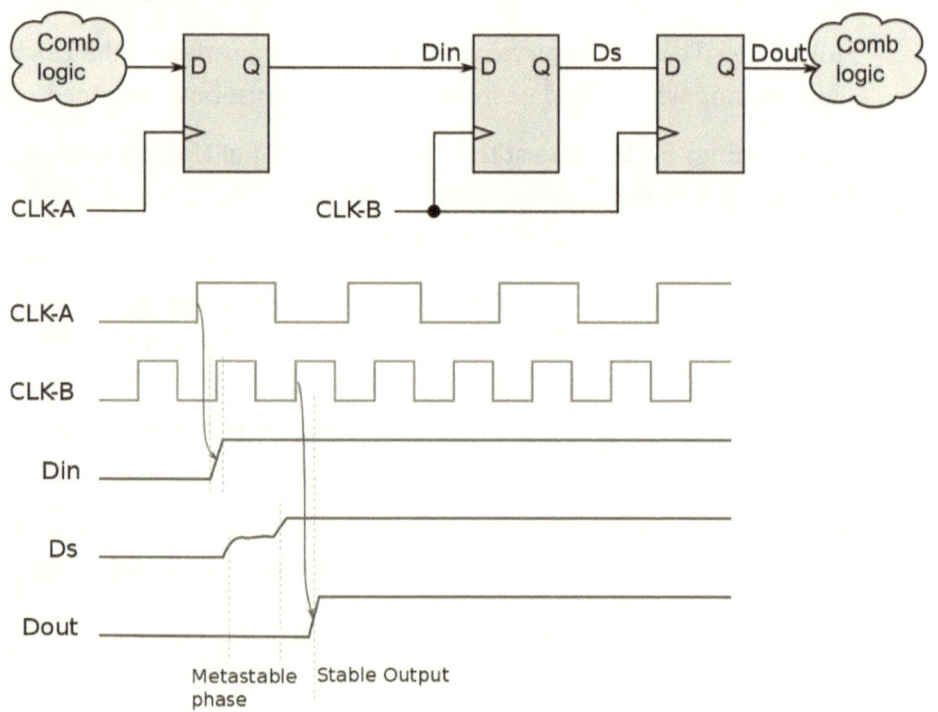

Figure 4.2: Metastability with two-clock input

Q3. Difference between Static Timing Analysis and Dynamic Timing Analysis?

Ans. **Static timing analysis** is the method of validating the timing performance of a design by checking all possible paths for timing violations under worst-case conditions. It considers the worst possible delay through each element but not the logical operation of the circuit.

Dynamic timing analysis verifies the functionality of the design by applying the input vectors and checking the correct output vectors.

Q4. Difference between Slack and Slew?

Ans. Slew has to do with transition time. It is the transition time of the signal, for the output voltage to go from 90% of Vmax to its 10% value.

This is not very closely related to Slack, which is the difference between Required arrival time of the signal to its actual arrival time.

Required time: The time within which data is required to arrive at some internal node of the design. The designer specifies this value by setting constraints.

Arrival Time: The time in which data arrives at the internal node. It incorporates all the net and logic delays in between the reference input point and the destination node, whereas Clock Skew can be compared with Slack. Clock skew is a phenomenon in synchronous digital circuit systems in which the same sourced clock signal arrives at different components at different times.

Q5. What is Clock Domain Crossing (CDC)?

Ans. A clock domain crossing occurs whenever data is transferred from a flop driven by one clock to a flop driven by another clock. The relationship between these clocks defines might leads to multiple issues for clock domain crossing. If the right techniques for synchronization are not applied, the circuit can go either in metastability or data loss. These days Static tools are smart enough to find such problematic crossings at the RTL level, and the user can correct very early.

Q6. How CDC can be avoided or removed?

Ans. There are different ways of avoiding and removing the CDC:

1. Conventional two flip-flop synchronizer

In general, a conventional two flip-flop synchronizer is used for synchronizing a single bit level signal. Flip-flop A and B1 are operating in the asynchronous clock domain. There is the probability that while sampling the input B1-d by flip flop B1 in CLOCK_B clock domain, output

B1-q may go into a metastable state. But during the one clock cycle period of CLOCK_B clock, output B1-q may settle to some stable value. The output of flop B2 can go to metastable if B1 does not settle to stable value during one clock cycle, but the probability for B2 to be metastable for a complete destination clock cycle is very close to zero.

A greater number of flop stages may be used if the frequency is too high as it will help in reducing the probability of synchronizer output to remain in a metastable state.

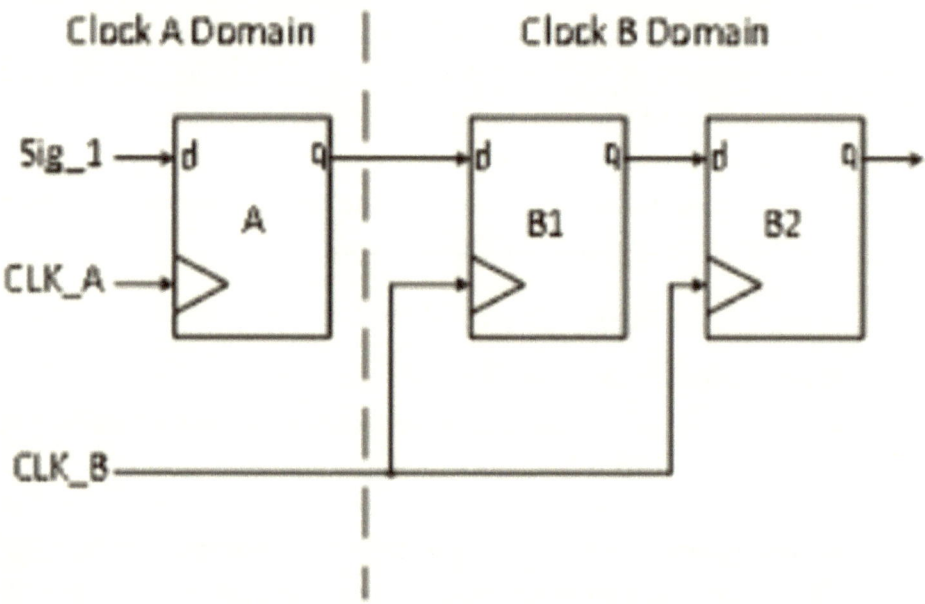

Figure 4.3: Conventional 2FF synchronizer

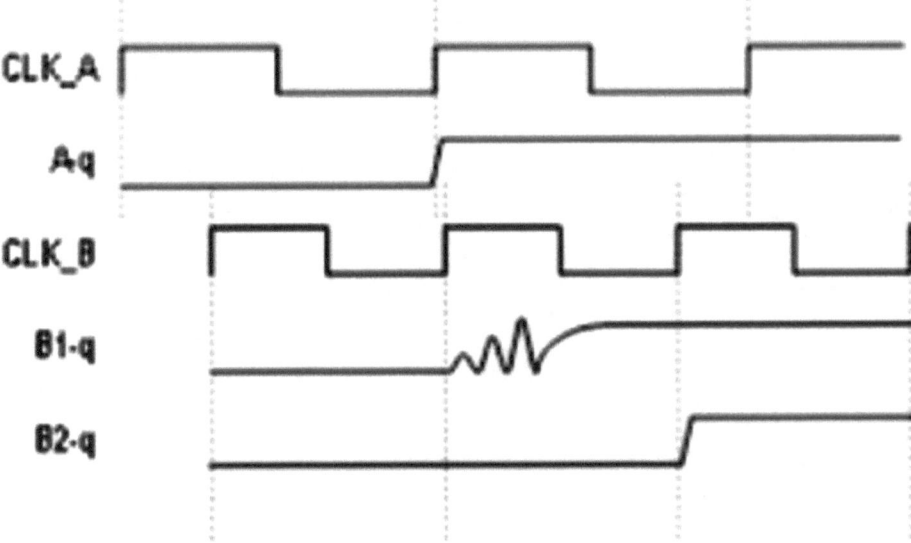

Figure 4.4: Timing for Conventional 2FF synchronizer

2. Toggle synchronizer

Toggle synchronizer is used to synchronize a pulse generating in source clock domain to destination clock domain. A pulse cannot be synchronized directly using 2 FF synchronizers. While synchronizing from fast clock domain to slow clock domain using 2 FF synchronizers, the pulse can be skipped which can cause the loss of pulse detection & hence subsequent circuit which depends upon it, may not function properly.

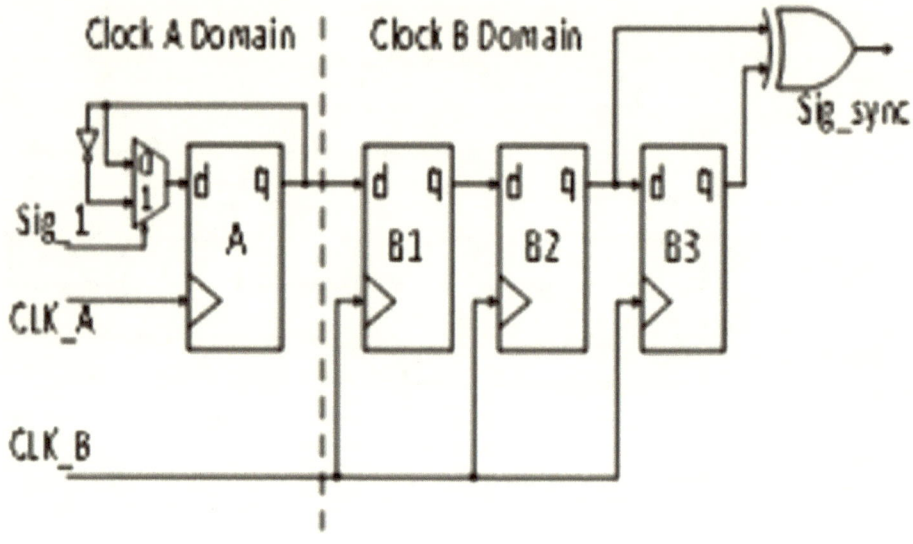

Figure 4.5: Toggle synchronizer

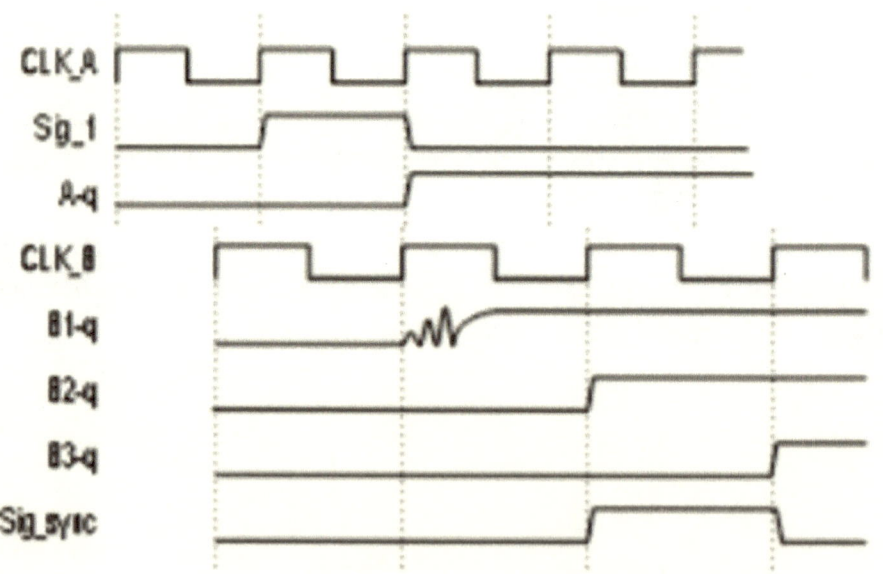

Figure 4.6: Timing for Toggle synchronizer

3. Handshake based pulse synchronizer

In handshake-based pulse synchronizer, synchronization of a pulse generated into the source clock domain is guaranteed into the destination clock domain by providing an acknowledgement. There is one restriction in pulse synchronizer that back to back (one clock gap) pulses can not be handled. To make sure the next generated pulse in the source clock domain gets transferred and synchronized in the destination clock domain, the handshake-based pulse synchronizer generates a "Busy" signal by ORing A1 and A3 flip-flop outputs. Thus, the logic generating the pulse shall not generate another pulse until the busy signal is asserted.

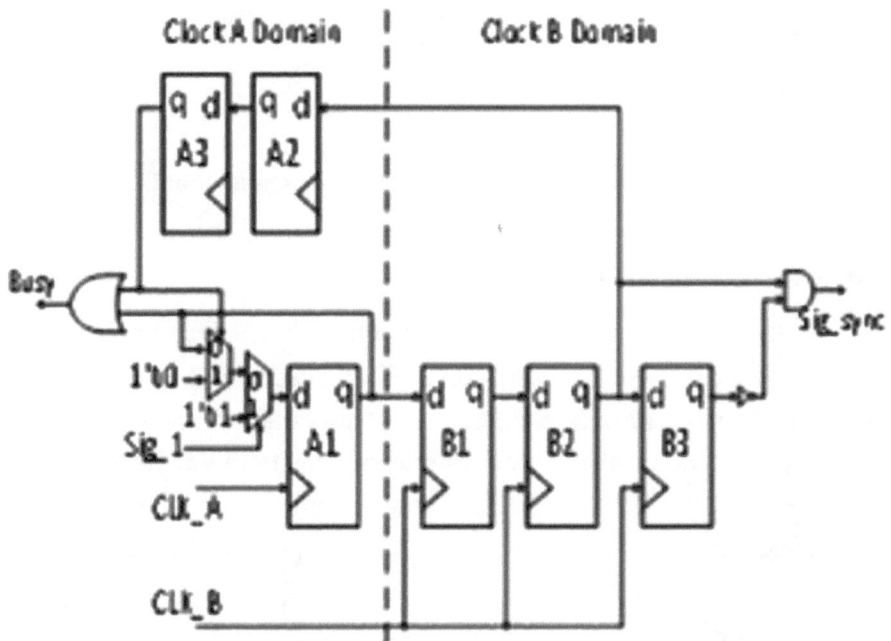

Figure 4.7: Handshake based pulse synchronizer

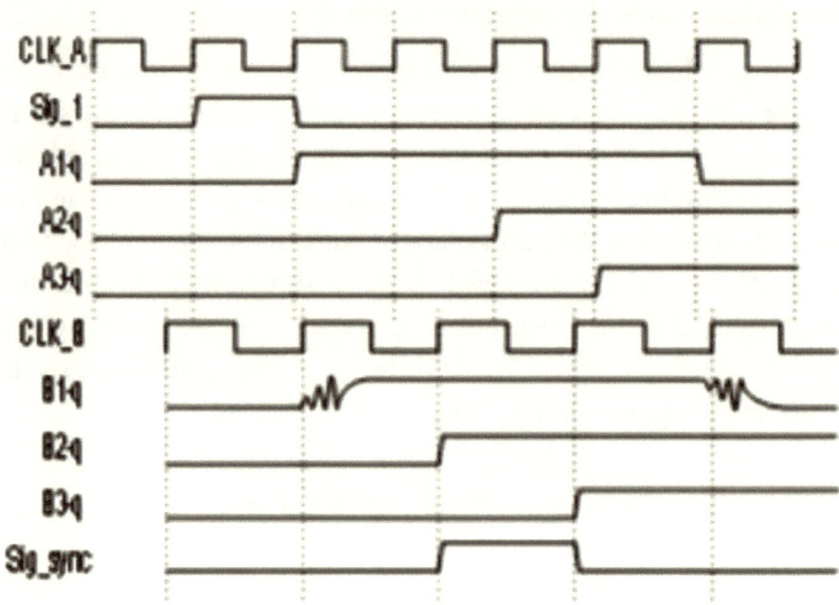

Figure 4.8: Timing for handshake-based pulse synchronizer

4. Gray encoding for multi bits signal

When multi-bit signals are synchronised with two flip flop synchronizers, each bit is synchronised using separate 2-FF synchronizer. Metastability can cause a flip flop to settle down either to the true value or false value. So, the output of every synchronizer may not settle to correct value at the same clock. This causes data incoherency. To synchronize multi-bit signal using two flip flop synchronizer method, only a single bit change must be guaranteed at a clock cycle. This can be achieved by gray encoding. So, for example, in asynchronous FIFO design, when we synchronise read pointer value after converting to gray value in write clock domain using 2-FF synchronizer, there is a possibility of metastability. As there is the only one-bit change in the gray encoding, so even if there is metastability when clock crossing, the gray counter value will be previous value. For example, read pointer (gray counter) value is changing from 0110 to 0111 and synchronised with write clock then due to metastability (if it occurs) possibility is read pointer remains 0110. Now, suppose earlier FIFO Full Flag was high at reading gray counter value 0110, then FIFO Full will remain high for one more clock cycle, but this won't cause an issue, because in the next clock cycle the read pointer value will become 0111 and FIFO full flag will get de-asserted. If instead of gray counter binary counter is taken from one clock domain to another through two flip flop synchronizers, then the multi-bit change could cause unpredicted recovery of different bits after metastability (e.g. value change from "1001" to "1010"). The recovered read or write pointer value could be erroneous, causing wrong Flag (FIFO full or FIFO empty) generation.

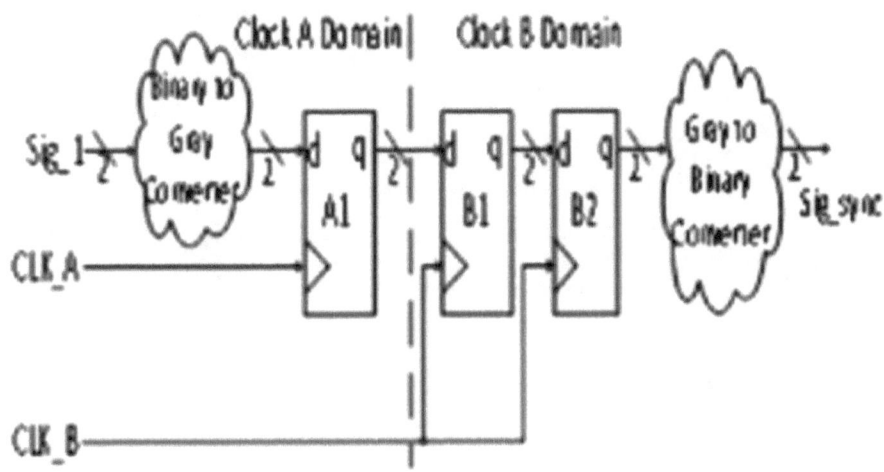

Figure 4.9: Gray encoding for multi-bit signal

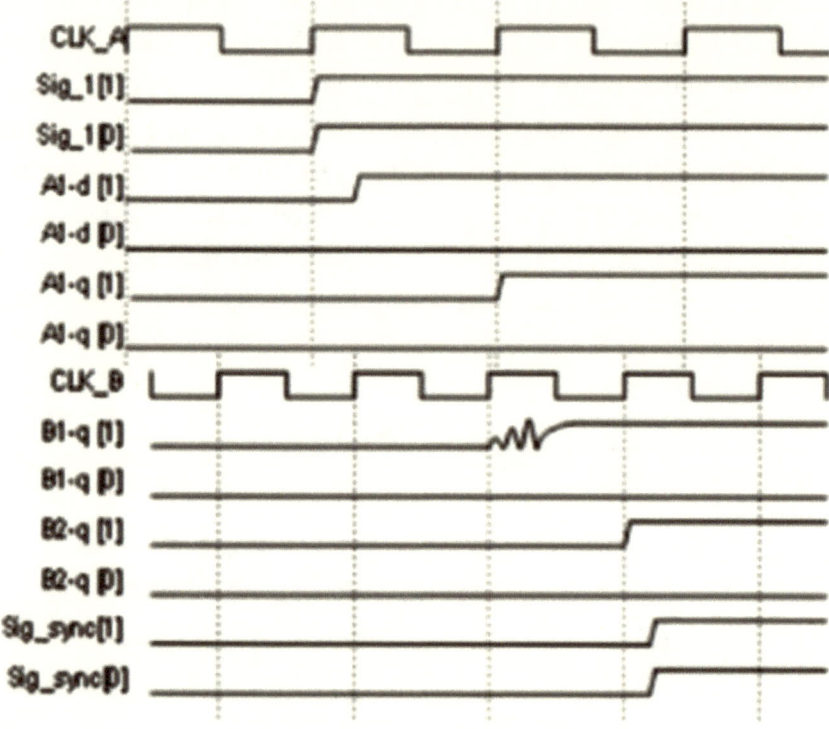

Figure 4.10: Timing for gray encoding for multi-bit signal

5. Recirculation mux synchronization

For isolated data and where multiple bits can transit at the same time, Recirculation mux synchronization technique, to synchronize data, a control pulse is generated in source clock domain when data is available at source flop. Control Pulse is then synchronized using two flip flop synchronizer or pulse synchronizer (Toggle or Handshake) depending on clock ratio between source and destination domain. The synchronized control pulse is used to sample the data on the bus in the destination domain. Data should be stable until it is sampled in the destination clock domain.

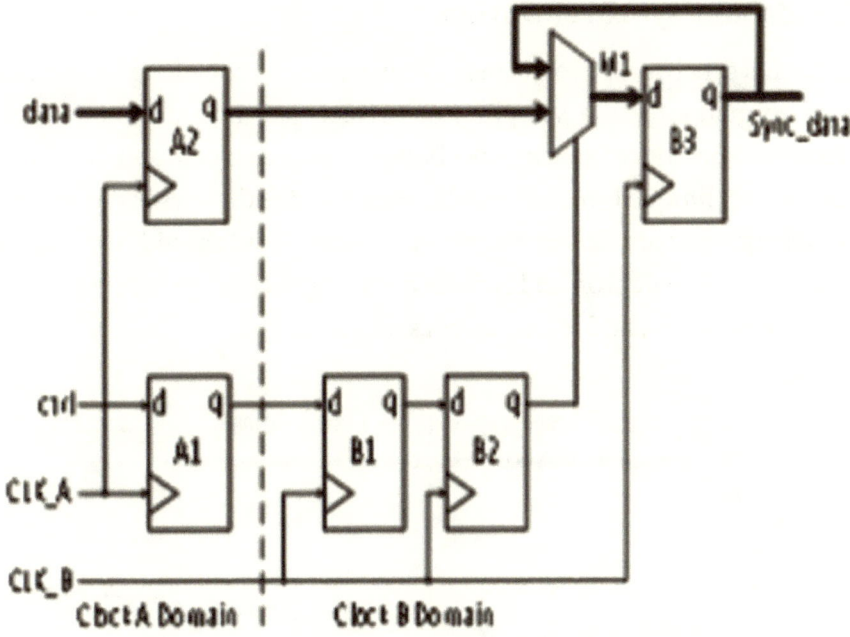

Figure 4.11: Recirculation mux synchronizer

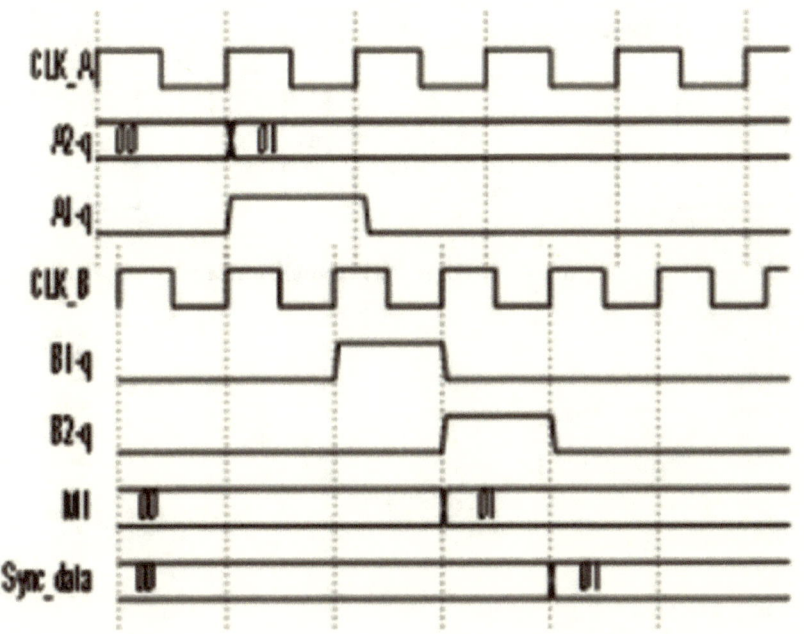

Figure 4.12: Timing for Recirculation mux synchronizer

6. Handshake synchronization

Ans: In this synchronization scheme request and acknowledge mechanism is used to guarantee the sampling of correct data into destination clock domain irrespective of clock ratio between source clock and destination clock. This technique is mainly used to synchronize the vector signal, which is not changing continuously or very frequently. Data should remain stable on the bus until synchronized Acknowledge signal (A2-q) is received from the destination side, and it (A2-q) goes low.

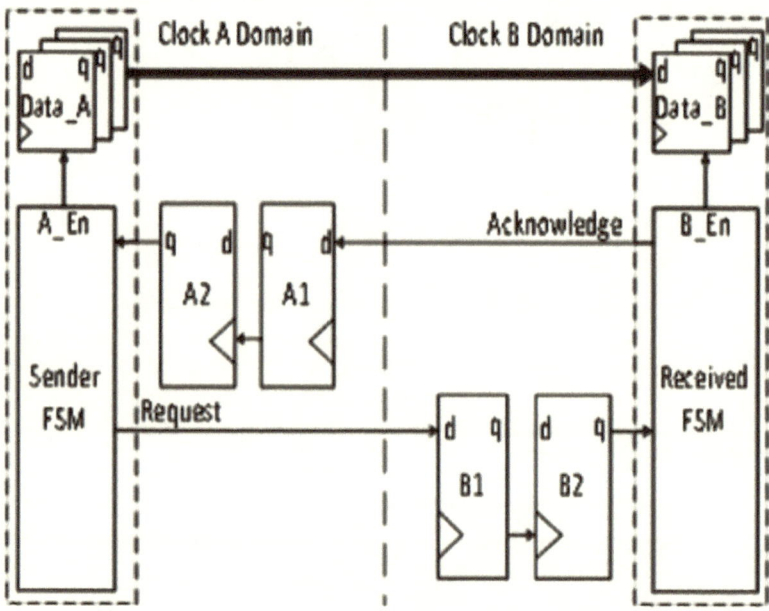

Figure 4.13: Handshake synchronizer

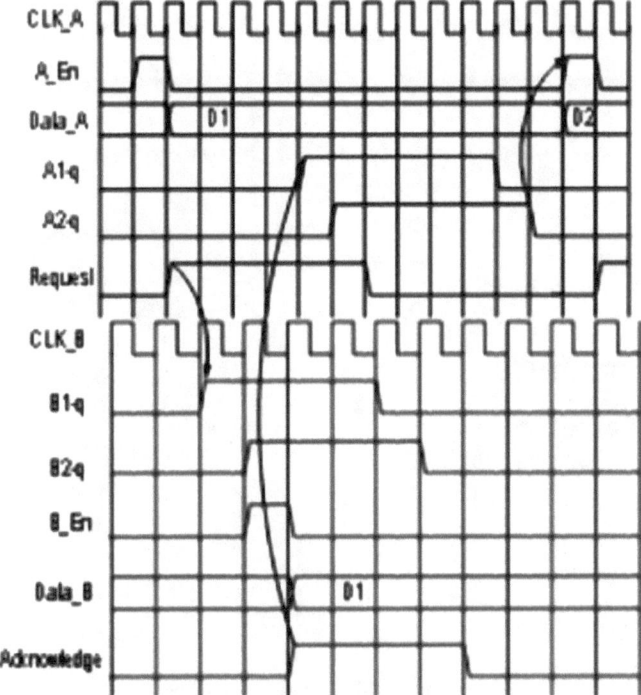

Figure 4.14: Timing for Handshake Synchronizer

7. Asynchronous FIFO synchronization

FIFO is the best way to synchronize continuously changing vector data between two asynchronous clock domains. Asynchronous FIFO synchronizer offers a solution for transferring vector signal across clock domains without risking metastability and coherency problems.

In Asynchronous FIFO design, FIFO provides full synchronization independent of clock frequency.

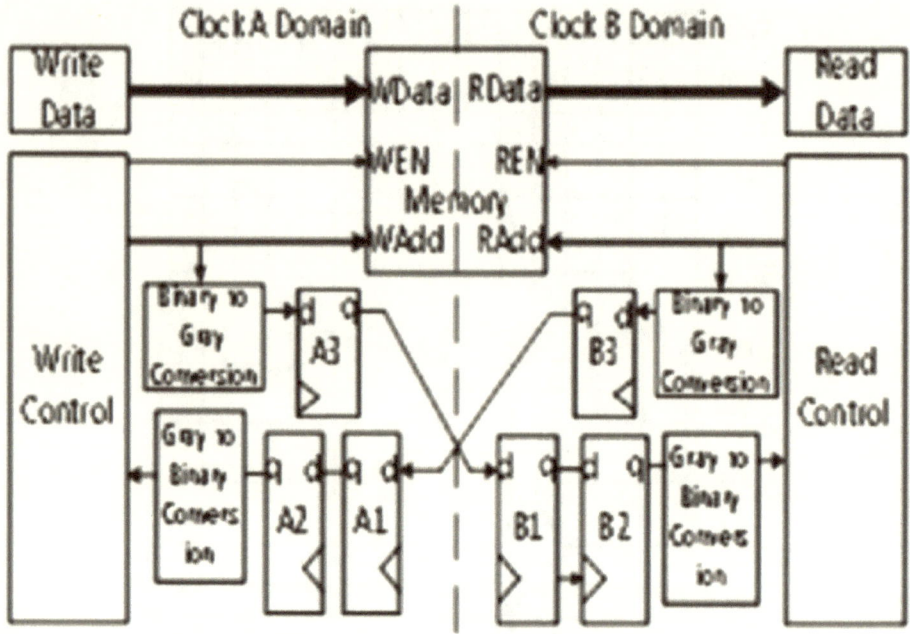

Figure 4.15: FIFO synchronizer

Q7. Difference between one hot and binary encoding?

Ans. Common classifications used to describe the state encoding of an FSM are Binary (or highly encoded) and One hot. A binary-encoded FSM design only requires as many flip-flops as are needed to uniquely encode the number of states in the state machine. The actual number of flip-flops required is equal to the ceiling of the log-base-2 of the number of states in the FSM.

A one hot FSM design requires a flip-flop for each state in the design, and only one flip-flop (the flip-flop representing the current or "hot" state) is set at a time in a one hot FSM design. For a state machine with 9- 16 states, a binary FSM only requires four flip-flops while a one hot FSM requires a flip-flop for each state in the design. FPGA vendors frequently recommend using a one-hot state encoding style because flip-flops are plentiful in an FPGA and the combinational logic required to implement a one hot FSM design is typically smaller than most binary encoding styles. Since FPGA performance is typically related to the combinational logic size of the FPGA design, one hot FSMs typically run faster than a binary encoded FSM with larger combinational logic blocks.

Q8. Explain ten ways for fixing setup and hold violations?

Ans. 8 ways to fix setup violations:

Setup violations are essentially where the data path is too slow compared to the clock speed at the capture flip flop with that in mind, and there are several things a designer can do to fix the setup violations.

Method 1: Reduce the amount of buffering in the path

- It will reduce the cell delay but increase the wire delay. So if we can reduce more cell delay in comparison to wire delay, the effective stage delay decreases.

Method 2: Replace buffers with two inverters place farther apart

- I am adding two inverters in a place of 1 buffer, reducing the overall stage delay.
- Adding inverter decreases transition time two times than the existing buffer gate. Due to the RC delay of the wire (interconnect delay decreases).
- As such cell delay of 1 buffer gate =cell delay of 2 inverter gate.
- So, stage delay (cell delay+wire delay) in the case of a single buffer<stage delay in the case of 2 inverters in the same path.

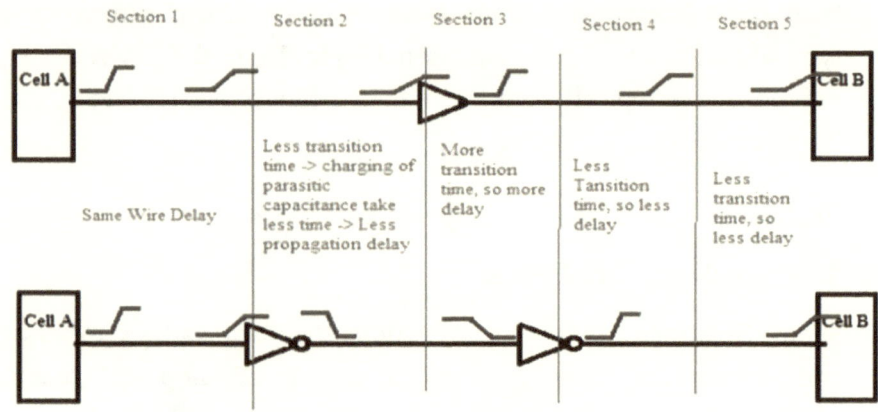

If, Cell delay of Buffer = 2*(Cell delay of NOT gate)
and Driving strength of Buffer = Driving Strength of NOT gate.
Stage Delay (In case of 1 Buffer) > Stage Delay (In case of 2 NOT gate)

Figure 4.16:

Method 3: HVT Swap. Means HVT cells into SVT/RVT or LVT.

- ❖ Low Vt decreases the transition time and propagation delay increases.
- ❖ HVT/NVT/LVT type cells have the same size and pin position. In both leakage current and speed LVT>NVT>HVT.So replace HVT with NVT or LVT will speed up the without disturb layout.
- ❖ Negative effect: Leakage current /Power also increases.

Method 4: Increase Driver size or say increase Driver strength (also known as upsizing the cell)

Note: Normally larger cell has higher speed. But some special cell may have larger cell slower than a normal cell. Check the technology library timings table to find out these special cells — the increasing driver commonly used in setup fix.

- ❖ Negative effect: Higher power consumption and more area used in the layout.

Static Timing Analysis ▪ 127

- ❖ The basic layout technique for reading the gate delay consists in connecting MOS devices in parallel.
- ❖ The equivalent width of the resulting MOS devices is the sum of each elementary gate width. Both nMOS and pMOS devices are designed using parallel elementary devices.
- ❖ The x1 inverter has the minimum size and is targeted for low speed, low power operations.
- ❖ The x2 inverter uses two devices x1 inverters, in II. The resulting circuit is an inverter with current capabilities. The output capacitance may be charged and discharged twice as fast as for the basic inverter, because of the Ron resistance of the Mos device is divided by two. The price to pay its a higher power consumption.
- ❖ The equivalent Ron resistance of the x4 inverters id divided by four.

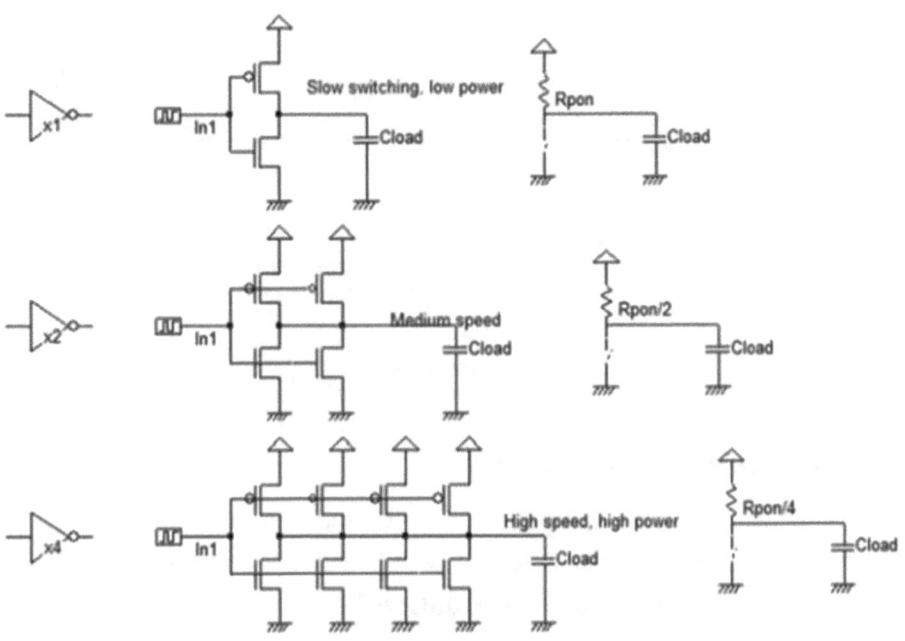

Figure 4.17:

Method 5: Insert Buffers

* Sometimes we insert the buffer to decrease the overall delay in case of log wire.
* Inserting buffer decreases the transition time, which decrease the wire delay.
* If the amount of the wire delay increases due to decreasing of transition time >cell delay of the buffer, overall delay decreases.
* Negative effect: Area will increase and increase in power consumption.

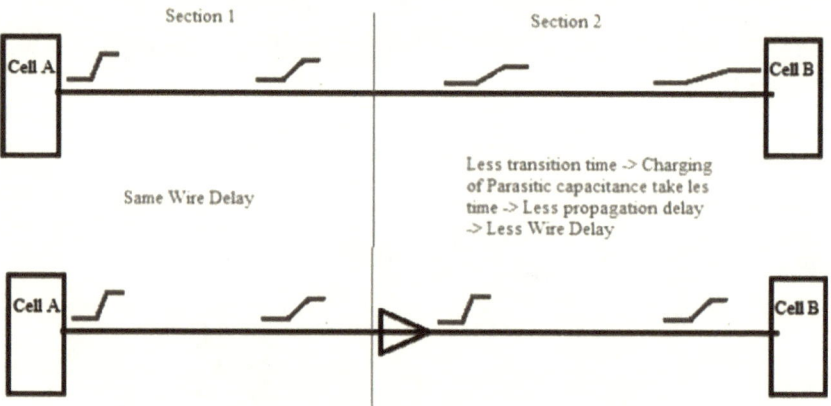

Wire Delay of Section 2 in case of Buffer < Wire Delay of Section2 with out Buffer
If, Cell Delay of Buffer < Delta Wire Delay of Section 2
Then Adding Buffer Decrease the over all Stage Delay

Figure 4.18:

Method 6: Inserting Repeaters

* Long-distance routing means a high RC loading due to a series of RC delays. A good alternative is to use repeater by splitting the line into several pieces. **Why can this solution be better in terms of delay? Because the gate delay is quite small compared to the RC delay.**

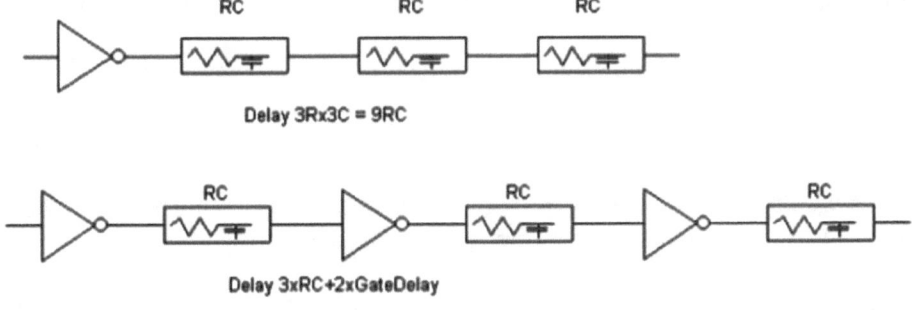

Figure 4.19:

- In the case of interconnect driver by a simple inverter, the propagation delay becomes

  Tdelay=tgate+nR*nC=tgate+n2RC

- If two repeaters are inserted, the delay becomes

  Tdelay=tgate(delay of inverters)+2 tgate(delay of repeater) +3RC= 3tgate + 3RC

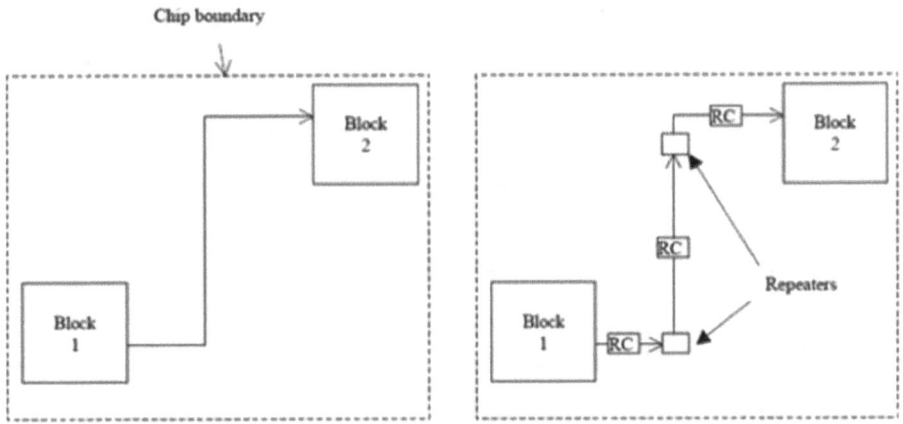

Figure 4.20: Inserting repeaters in long lines

Method 7: Adjust cell position in the layout

* Let's assume there are two gates (GATE A and GATE B) separated by 100um. There is another GATE C place at a distance of 900 um from GATE A.

* If we reposition GATE C at 500 um from GATE A (centre of GATE A and B), overall between GATE A and GATE B decreases.

Note: **The placement in the layout may prevent such movement. Always use layout viewer to check if there are any spare space to move the critical cell to an optimal location.**

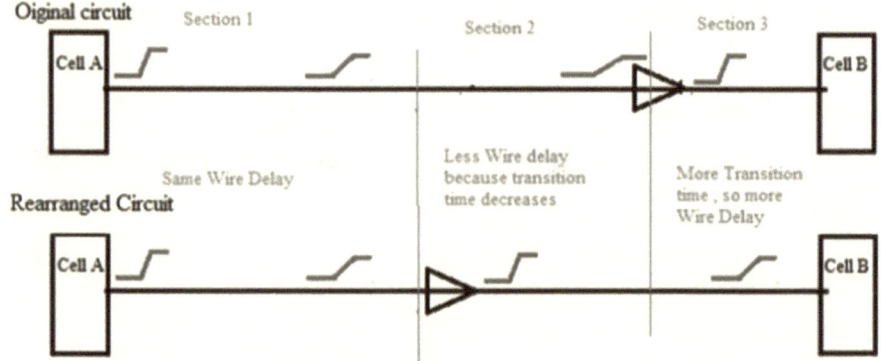

If you will see closely then you come to know that transition time is very bad in section 2 (for original circuit) in comparison to the transition time in section 3 (rearranged circuit). That mean, **Delta wire delay in section 2 > Delta wire delay in section 3**

Figure 4.21:

Method 8: Clock Skew

* By delaying the clock to the endpoint can relax the timing of the path, but you must make sure the downstream paths are not the critical path.

2 Ways to fix hold violations

Hold violation is the opposite of setup violation. Hold violation happen when data is too fast compared to clock speed. For fixing the hold violations, the delay should be increased in the data path.

Note: Hold violations are critical and on priority basis in comparison are not fixed before the chip is made.

Method 9: By adding delays

- Adding buffer/inverter pairs/delay cells to the data path helps to fix the hold violation.

   Note: The hold violation path may have its start point or endpoint in other setup violation paths. So we have to take extra care before adding to the buffer delay.

   E.g., IF the endpoint of hold violation path has setup violation concerning some other path, insert buffer/ delay near the start point of hold violation path. Else the setup violation increases in another path.

- If the start point of hold violation path has set up violation concerning some other paths, insert the buffer/delay near the endpoint of hold violation path. Else setup violation increase in another path.

- I am sure you may be asking what is this and why?

- Below figure and explanation can help you to understand this.

- **From the below figure, you can also conclude that don't add buffer/delay in the common segment of 2 paths (where one path has hold violation and other setup violation)**

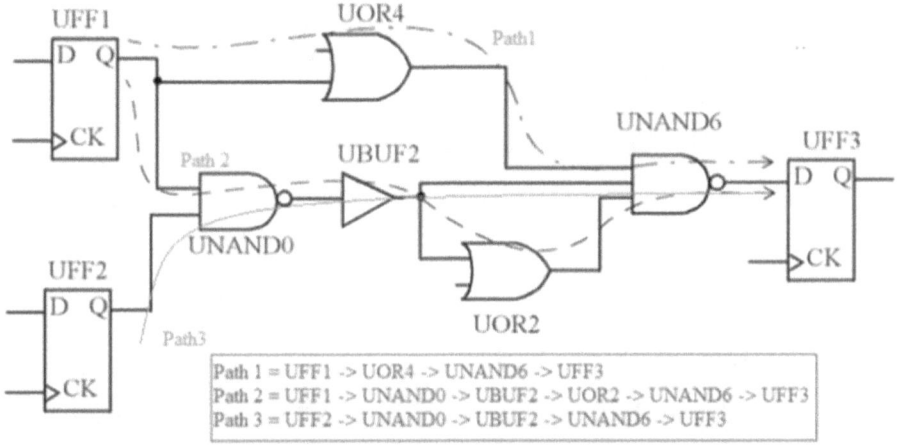

Figure 4.22:

Violation Paths		For Fixing the Hold Violation	
Setup	Hold	Do's	Don'ts
Path1	Path2	Add delay anywhere in Path2 between UNAND0 & UNAND6	Don't add delay between UNAND6 & UFF3 (Common section of both paths)
Path1	Path3	Add delay anywhere in between UFF2 & UNAND6	Don't add delay between UNAND6 & UFF3 (Common section of both paths)
Path2	Path3	Add delay before UNAND0 in path3. Can add a delay in path3 between UBUF2 & UNAND6	Don't add a delay between UNAND0 & UNAND6 (Common section of both paths)

Method 10: Decreasing the size of certain cells in the data path.

❖ It is better to reduce the cells closer to the capture flip flop because there is less likelihood of affecting other paths and causing new errors.

Q9. What are the timing paths?

Ans. Timing paths can be divided as per the type of signals (e.g. clock signal, data signal etc.)

Type of paths for timing analysis

❖ **Data Path**
❖ **Clock Path**
❖ **Clock Gating Path**
❖ **Asynchronous Path**

Each timing path has a "Start Point" and an "Endpoint". Definition of Start Point and End Point vary as per the type of the timing path. E.g. for the Data Path- A starting point is a place in the design where a clock edge launches data. The data is propagated through combinational logic in the path and then captured at the endpoint by another clock edge.

Start point and End Point are different for each type of paths. It's very important to understand this clearly to understand and analyzing the Timing analysis report and fixing the timing violation.

Data Path

❖ Start Point

1. The input port of the design (because the input data can be launched from some external source).
2. Clock pin of the flip-flop/latch/memory(sequential cell).

❖ End Point

1. Data input pin of the flip-flop/latch/memory(sequential cell).
2. An output port of the design (because the output data can be captured by some external sink)

Clock Path

❖ Start Point

1. Clock Input port

❖ End Point

1. Clock pin of the flip-flop/latch/memory (sequential cell).

Clock Gating Path

❖ Start Point

1. The input port of the design.

❖ End Point

1. The input port of clock-gating element.

Asynchronous Path

❖ Start Point
   1. The input port of the design
❖ End Point
   1. Set/Reset/Clear pin of the flip-flop/latch/memory(sequential cell).

Q10. Briefly describe the different types of timing paths?

Ans. Data Paths:

If we use all the combination of the two types of the starting point and two types of endpoint, we can say that there are four types of timing paths based on start and endpoint.

❖ Input pin/port to Register (flip-flop)

❖ Input pin/port to Output pin/port

❖ Register (flip-flop) to Register (flip-flop)

❖ Register (flip-flop) to Output pin/port

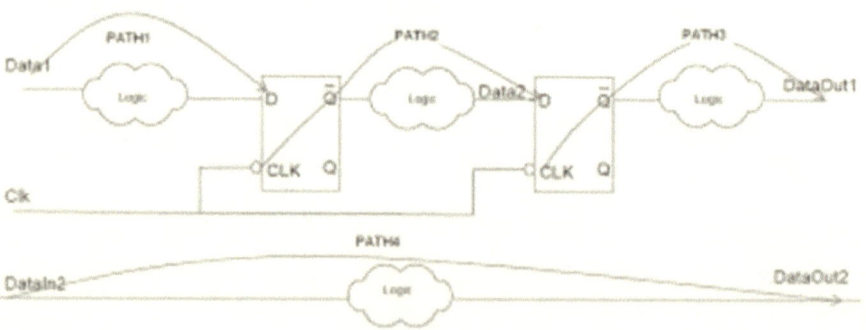

Figure 4.23:

PATH1 - starts at an input port and ends at the data input of a sequential element(Input port to Register).

PATH2- starts at the clock pin of a sequential element and ends at the data input of a sequential element(Register to Register).

PATH3- starts at the clock pin of a sequential element and end at an output port(Register to Output port).

PATH4- starts at an input port and ends at an output port(Input port to Output port).

Clock Paths:

Please check the following figure:

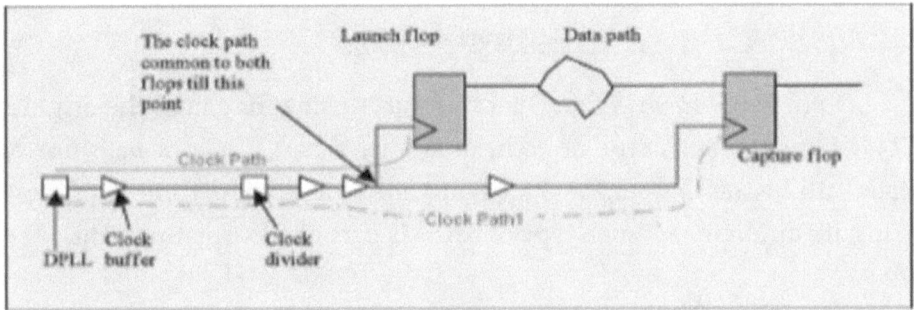

Figure 4.24:

In the above fig, it's very clear that for clock path that starts from the input port/pin of the design which is specific for the clock input and the endpoint is the clock pin of the sequential element. In between the start point and the endpoint, there may be lots of Buffers/Inverters/clock divider.

Clock Gating Path:

Clock Path may be passed through a "gated element" to achieve an additional advantage. In this case, characteristics and definitions of the clock change accordingly. We call this type of clock path as "gated clock path".

As in the following fig, you can see that

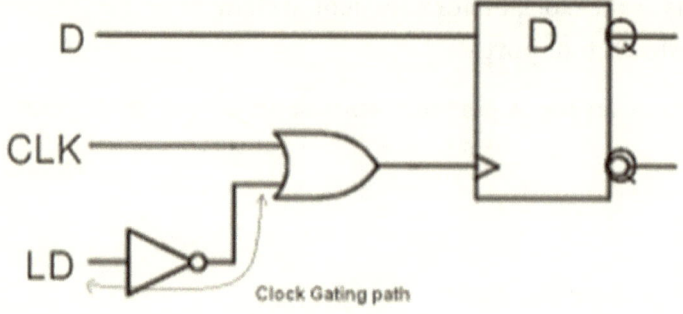

Figure 4.24:

LD pin is not a part of any clock, but it is using for gating the original CLOCK signal. Such type of paths is neither a part of Clock path nor of Data Path because as per the Start Point and End Point definition of these paths, it's different. So, such type of paths is part of Clock gating path.

Asynchronous Path:

A path from an input port to an asynchronous set or clear pin of a sequential element.

See the following fig for understanding.

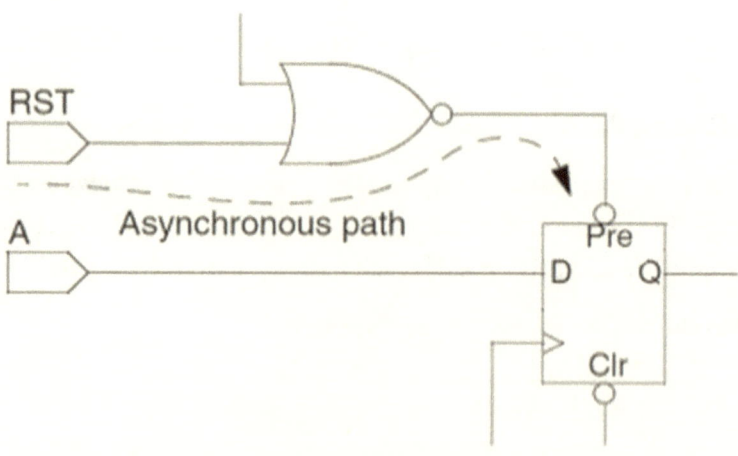

Figure 4.25:

As you know that the functionality of the set/reset pin is independent of the clock edge. It's level-triggered pins and can start functioning at any time of data. So, in another way, we can say that this path is not synchronous with the rest of the circuit and that's the reason we are saying such type of path an Asynchronous path.

Other types of Paths:

There are a few more types of a path which we usually use during timing analysis reports. Those are a subset of paths mentioned above with some specific characteristics. Since we are discussing the timing paths, so it will be good if we will discuss those here also.

Few names are

- Critical path
- False Path
- Multi-cycle path
- Single-cycle path
- Launch Path
- Capture Path
- Longest Path (also known as Worst Path, Late Path, Max Path, Maximum Delay Path)
- Shortest Path (also known as Best Path, Early Path, Min Path, Minimum Delay Path)
- Q11: Explain Critical Path?
- Ans: In short, I can say that the path which creates Longest delay is the critical path.
- Critical paths are timing-sensitive functional paths, because of the timing of these paths is critical, no additional gates can be added to the path, to prevent increasing the delay of the critical path.

- Timing critical path are those paths that do not meet your timing. What normally happens is that after synthesis is the tool will give you several paths which have a negative slag. The first thing you would do is to make sure those paths are not false or multi-cycle since it that case you can ignore them.

Taking a typical example in a very simpler way,(the STA tool will add the delay contributed from all the logic connecting the Q output of one flop to the D input of the next (including the CLOCK->A of the first flop). And then compare it against the defined clock period of the CLOCK pins (assuming both flops are on the same clock, and taking into account the setup time of the second flop and the clock skew). This should be strictly less than the clock period defined for that clock. If the delay is less than the clock period, then the "path meets timing". If it is greater than the "path fails timing". The "critical path" is the out of all the possibilities that either exceeds its constraint by the largest amount or, if all paths pass, then the one that comes closest to failing.

Q12. Explain the false path?

Ans. **False Path:**

- Physically exist in the design, but those are logically/functionally incorrect path. Means no data is transferred from Start Point and to End Point. There may be several reasons for such path present in the design.

- Sometimes we have to explicitly define/create few false paths within the design — E.G. for setting a relationship between two Asynchronous Clocks.

- The goal in static timing analysis is to do timing analysis on all "true" timing paths, and these paths are excluded from timing analysis.

- Since false path are not exercised during normal circuit operation, they typically do not meet timing specification, considering false path during timing closure can result in timing violations and the procedure to fix would introduce unnecessary complexities in the design.

- There may be a few paths in your design, which are not critical for timing or masking other paths which are important for timing optimization or never occur within a normal situation. In such case, to increase the run time and improving the timing result, sometimes we must declare such path as a False path so that Timing analysis tool ignore these paths and so the proper analysis concerning other paths. Or During optimization doesn't concentrate over such paths. One example of this, e.g. A path between two multiplexed blocks that are never enabled at the same time. You can see the following picture for this.

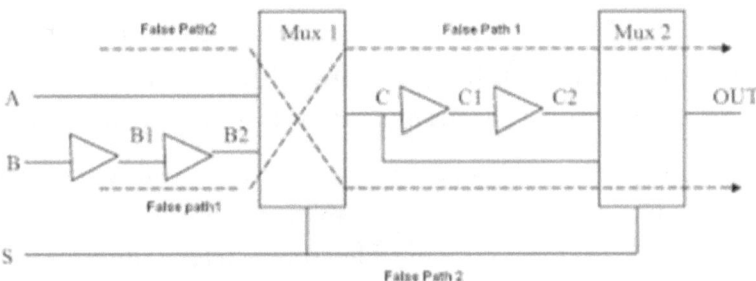

Figure 4.26:

Here you can see that False path1 and False Path 2 cannot occur at the same time, but during optimization, it can affect the timing of another path. So in such a scenario, we have to define one of the paths as a false path.

Same thing I can explain in another way (Note-Took snapshot from one of the forums). As we know that, not all paths that exist in a circuit are "real" timing paths. For example, let us assume that one of the primary inputs to the chip is a configuration input; on the board, it must be tied to either VCC or GND. Since this pin can never change, there are never any timing events in that signal. As a result, all STA paths that start at this start point are false, which the designer can do by telling the tool using a "false_path" directive. When told that the paths are false, the STA tool will not analyze it (and hence will not compare it to a constraint, so this path cannot fail), nor will a synthesis tool do any optimizations on that particular to make it faster. Synthesis tools try and improve paths until they "meet timing" - since the path is false, the synthesis tool has no work to do on this path.

Thus, a path should be declared false if the designer KNOWS that the path in question is not a real timing path, even though it looks like one to the STA tool. One must be very careful with declaring a path false. If you declare a path false, and there is ANY situation where it is a real path, then you have created the potential for a circuit to fail, and for the most part, you will not catch errors until the chip is on board, and (not) working. Typically, false paths exist

- From configuration inputs like the one described above
- From "test" inputs; inputs that are only used in the testing of the chip and are tied off in normal mode (however, there may still be some static timing constraints for the test mode of the chip)
- From asynchronous inputs to the chip (and you must have some form of synchronizing circuit on this input, and this is not an exhaustive list but covers most legitimate false paths).

So, we can say that false paths should NOT be derived from running the STA tool(or synthesis tool); they should be known by the designer as part of the definition of the circuit and constrained accordingly at the time of initial synthesis.

Q13. Explain Multicycle Path?

Ans. **Multicycle Path:**

- A multicycle path is a timing path that is designed to take more than one clock cycle for the data to propagate from the start point to the endpoint.

  A multi-cycle path is a path that is allowed multiple clock cycles for propagation. Again, it is a path that starts at a timing start point and ends at a timing endpoint. However, for a multi-cycle path, the normal constraint on this path is overridden to allow for the propagation to take multiple clocks.

In the simplest example, the start point, and endpoint are flops clocked by the same clock. The normal constraint is therefore applied by the definition of the clock; the sum of all delays from the CLOCK arrival at the first flop to the arrival at the D of the second clock should take no more than one clock period minus the setup time of the second flop and adjusted for clock skew.

By defining the path as a multicycle path, you can tell the synthesis or STA tool that the path has N clock cycles to propagate; so the timing check becomes " the propagation must be less than N x clock_period, minus the setup time and clock skew". N can be any number greater than 1.

Few examples are

❖ When you are doing clock crossing from two closely related clocks; From 30MHz clock to a 60MHz clock,

1. Assuming the two clocks are from the same clock source (i.e. one is the divided clock of the other), and the two clocks are in phase.

2. The normal constraint, in this case, is from the rising edge of the 30MHz clock to the nearest edge of the 60MHz clock, which is 16ns later. However, if you have a signal in the 60 MHz domain that indicates the phase of the 30MHz clock, you can design a circuit that allows for the full 33ns for the clock crossing, then the path from flop30-> to flop60 is an MCP(again with N=2).

3. The generation of the signal 30MHZ_is_low is not trivial since it must come from a flop which is clocked by the 60MHz clock but show the phase of the 30MHz clock.

❖ Another place would be when you have different parts of the design that run at different, but related frequencies. Again, consider a circuit that has some stuff running at 60Mhz and some running on a divided clock at 30MHz.

1. Instead of defining two clocks, you can only use the faster clock and have a clock enable that prevents the clocks in the slower domain from updating every other clock.

2. Then all the paths from the "30MHz" flop "30MHz" can be MCP.

3. This is often done since it is usually a good idea to keep the number of different clock domains to a minimum.

Q14. Explain the single cycle path?

Ans. **Single Cycle Path:**

A single-cycle path is a timing path that is designed to take only one clock cycle for the data to propagate from the start point to the endpoint.

Launch Path and Capture Path:

Both are inter-related, so I am describing both in one place. When a flip flop to flip-flop such as UFF1 To UFF3 is considered, one of the flip-flops launches the data, and others capture the data. So here UFF1 is referred to "launch Flip-flop", and UFF3 referred to "capture flip-flop";

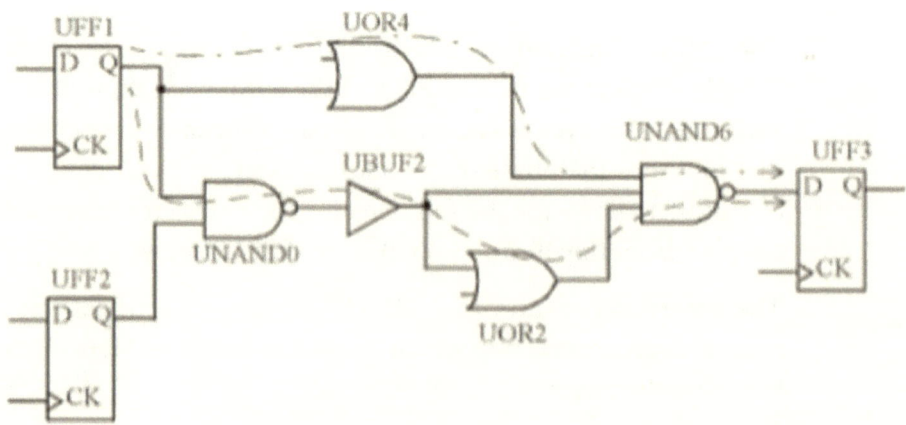

Figure 4.26:

These Launch and Capture terminologies are always referred to a flip-flop to flip-flop path. Means for this path (UFF1->UFF3), UFF1 Is launch flip=flop, and UFF3 is capture flip-flop. Now if there is any other path starting from UFF# and ends to some flip-flops(let's assume UFF4), then for that path UFF3 become launch flip-flop and UFF4 be as capture flip-flop.

The Name "Launch path" referred to a part of the clock path. Launch path is the launch clock path which is responsible for launching the data at launch flip flop. And Similarly, Capture path is also a part of the clock path. Capture path is capture clock path which is responsible for capturing the data at capture flip flop. This can be understood from the below figure :

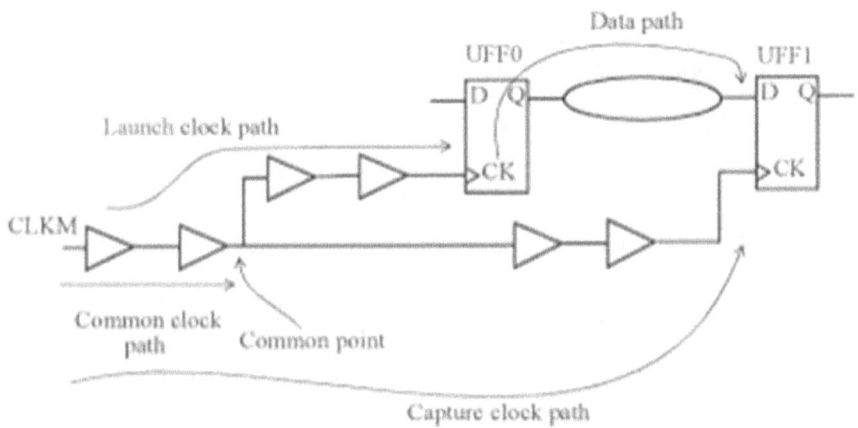

Figure 4.27:

Here UFF0 is referred to launch flip-flop and UFF1 as capture flip-flop for "Data path" between UFF0 to UFF1. So, the start point for this data path is UFF0/CK and endpoint is UFF1/D.

- ❖ Launch path and data path together constitute arrival time of data at the input of capture flip-flop.
- ❖ Capture clock period and its path delay together constitute the required time of data at the input of the capture register.

Note: It's very clear that capture and launch paths correspond to the Data path. Means same clock path can be a launch path for one data path and be a capture path for another datapath. Its will be clear by the following fig (source of Fig is from Synopsys)

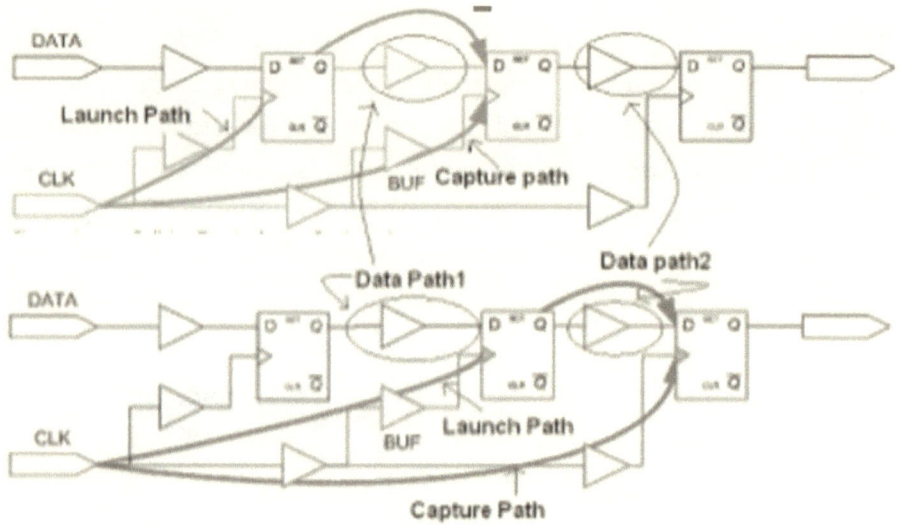

Figure 4.28:

Here you can see that for Data path1 the clock path through BUF cell is capture path but for Data path2 it's a Launch Path.

Longest and Shortest Path:

Between any 2 points, there can be many paths.

The longest path is the one that takes the longest time, and this is also called the worst path or late path or a max path.

The shortest path is the one that takes the shortest time; this is also called the best path or early path or a min path.

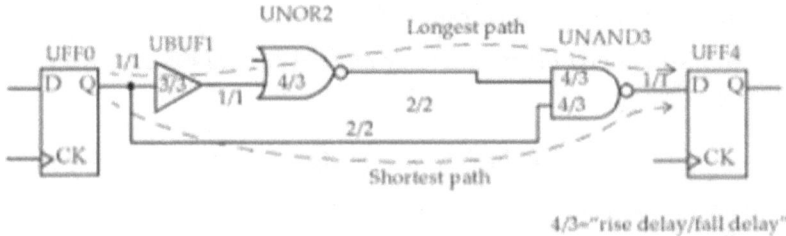

Figure 4.29:

In the above fig, the longest path between the two flip-flops is through the cells UBUF1, UNOR2 and UNAND3. The shortest path between the two flip-flops is through the cell UNAND3.

Q15. What is Maximum and Negative Borrow time?

Ans. **Maximum Borrow time** is the clock pulse width minus the library setup time of the latch. Usually to calculate the maximum allowable borrow time, start with clock pulse and then subtract clock latency, clock reconvergence pessimism removal, the library setup time of the endpoint latch.

Negative Borrow time: If the arrival time minus the clock edge is a negative number, the amount of time borrowing is negative (in another way, you can say that no borrowing). This amount is known as Negative Borrow time.

Q16. Explain the difference between latch based and a flop-based design?

Ans. The basic differences between the latch-based design and flipflop-based design is as follows:

- ❖ Edge-triggered flip-flops change states at the clock edges, whereas latches change states if the clock pin is enabled.

- ❖ The delay of a combinational logic path of a design using edge-triggered flip-flops cannot be longer than the clock period except for those specified as false paths and multiple cycle paths. So, the performance of a circuit is limited by the longest path of design.

❖ In latch-based design, the longer combinational path can be compensated by shorter path delays in the subsequent logic stages. So, far higher performance circuits designers are turning to latch based design.

It's true that in the latched based design, it's difficult to control the timing because of multi-phase clocks used and the lack of "hard" clock edges at which events must occur.

Q17. What is time borrowing or cycle stealing?

Ans. The technique of borrowing time from the shorter paths of the subsequent logic stages to the longer path is called **time borrowing** or **cycle stealing.**

Let's talk about this. Please see the following figure,

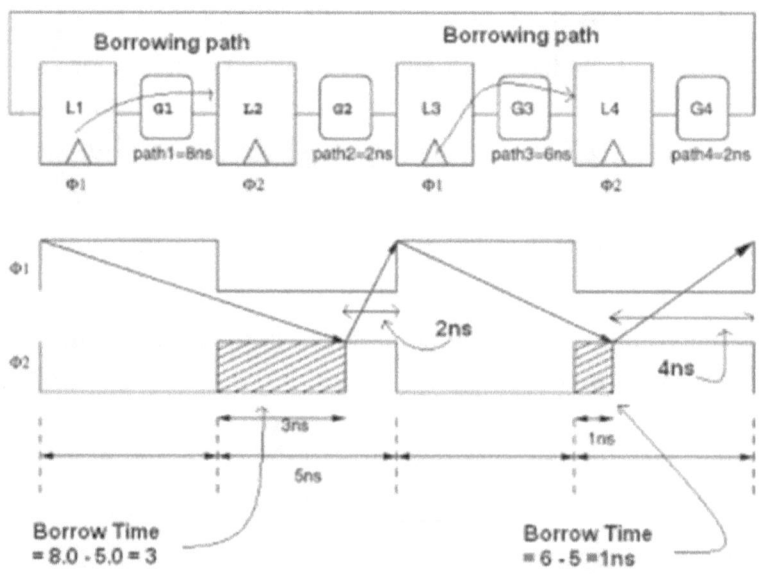

Figure 4.30:

There are four latches (positive level sensitive). PH1 controls L1 and L3, and PH2 controls L2 and L4. G1, G2, G3 and G4 are combinational logic paths. For now, assume a library setup time is zero for the latches and zero delays in latch data-path in the transparent mode.

Now, assume that if designs using edge-triggered flip-flops, the clock period must be at least 8ns because the longest path in G1 is 8ns. Now, as the clock pulse is 5ns, there is a violation at L2. On the other hand, if the design uses latches, L2 latch is transparent for another 5ns, and since the eighth (8th) ns is within the enabled period of L2, the signal along path1 can pass through L2 and continue path2. Since the delay along the path is 2ns, which is short enough to compensate for the overdue delay of path1, this design will work properly. In other words, we can say that path1 can sometimes borrow(3ns) from path2. Since the sim of path1 and path2 is 10ns, which is required the time of L3, there will be no violation in either of the latches.

For the same reason, path3 can borrow some time (1ns) from path4 without any timing violation.

Note: latch-based design completes the execution of the four logic stages in 20ns, whereas an edge-triggered based design needs 32ns.

Let see this in a more complex design. It's self-explanatory

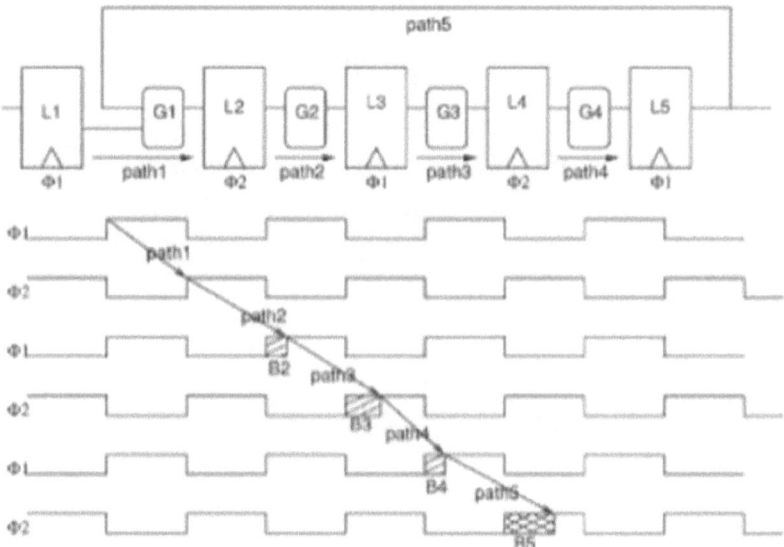

Figure 4.31:

Just wanted to convey here that this Timing borrowing can be multistage. Means we can easily say that for a latched based design, each executing path must start at a time when its driving latch is enabled, and end at a time when its driven latch is enabled.

Few Important things:

- ❖ Time borrowing occurs within the same cycle. Means launching and capturing latches be using the same phase of the same clock. When the clocks of the launching and capturing latches are out of phase, time borrowing is not to happen. Usually, it was disabled by EDA tools.

- ❖ Time borrowing typically only affects setup slack calculation since time borrowing slows data arrival times. Since hold time slack calculation, uses fastest data, time borrowing typically does not affect hold slack calculation.

Q18. How to calculate the maximum clock frequency?

Ans.

Example1: Multiple FF's Sequential Circuit

In typical sequential circuit design, there are often millions of flip-flop to flip-flop paths that need to be considered in the calculation the maximum clock frequency. This frequency must be determined by locating the longest path among all the flip-flop paths in the circuit. Consider the following circuit.

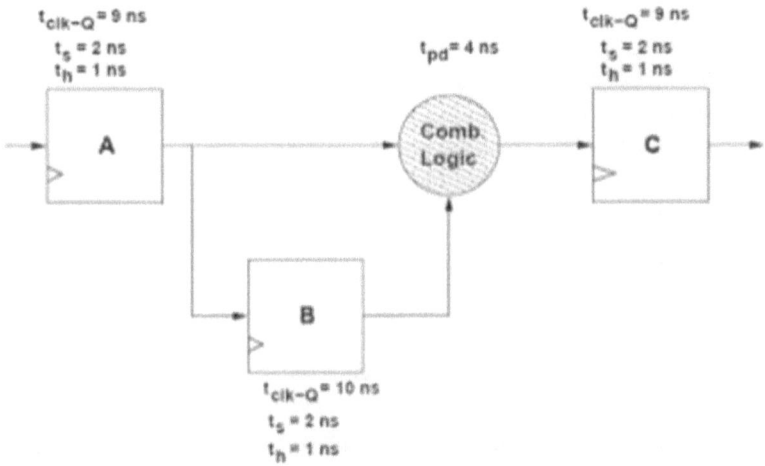

Figure 4.32:

There are three flip-flops to flop-flop paths (flop A to flop B, flop A to flop C, flop B to flop C).

- TAB= tClock-Q(A)+ts(B)=9ns+2ns=11ns
- TAC= tClock-Q(A)+tpd(Z)+ts(C)=9ns+4ns+2ns=15ns
- TBC=tClock-Q(B)+tpd(Z)+ts(C)=10ns+4ns+2ns=16ns

Since the TBC is the largest of the path delays, the minimum clock period for the Tmin=16ns and the maximum clock frequency is 1/Tmin.

Example 2: Circuit with min and max delay specification

Let's consider the following circuit. Now this circuit is like normal FF circuitry, only differences are

- Every specification has two values (Min and Max).
- There is a combinational circuit in the clock path also.

150 • Gateway to VLSI

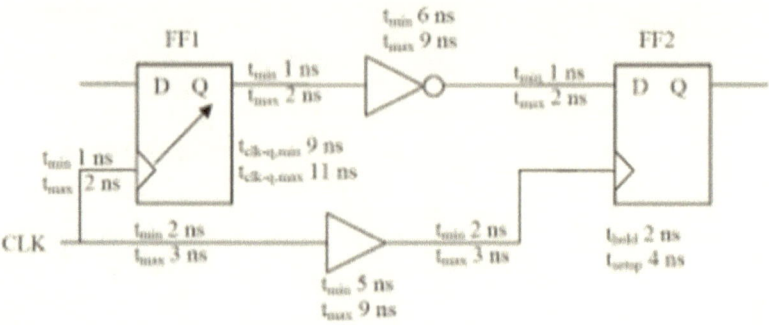

Figure 4.33:

Now let's understand the flow/circuit once again.

* Every interconnect wire also has some delay, so you see the clock CLOCK will take some time to reach the clock pin of FF1.
* That's means concerning the original clock edge(let's assume at 0 ns), clock edge will take minimum 1ns and maximum 2ns to reach the clock pin of FF1.
* So similarly, if we calculate the total minimum delay and maximum delay.
  1. In data path: max delay =(2+11+2+9+2)ns=26ns
  2. In data path: min delay=(1+9+1+6+1)ns=18ns
  3. In clock path: max delay=(3+9+3) ns=15ns
  4. In clock path:min delay=(2+5+2)ns=9ns
* So, for minimum clock period, we want to make sure that at FF2, data should be present at least "setup" time before positive clock edge(if it's a positive edge-triggered flipflop) at the FF2.
  1. So, clock edge can reach at the FF2 after 9ns/15ns (min/max) with the reference of the original clock edge.
  2. And data will take time 18ns/26ns(min/max) with the reference of the original clock edge.
  3. So, clock period in all the 4 combinations are

a. Clock period(T1)=(Max data path delay)-(max clock path delay)+tsetup=26-15+4=15ns.

b. Clock period(T2)=(Min data path delay)-(max clock path delay)+tsetup=18-15+4=7ns

c. Clock period(T3)=(Max data path delay)-(min clock path delay)+tsetup=26-9+4=21ns

d. Clock period(T4)=(Min data path delay)-(min clock path delay)+tsetup=18-9+4=11ns

❖ Since we want that this circuit should work in the entire scenario (all combination of data and clock path delay), so we must calculate the period based on that.

1. Now if you will see all the above clock period, you can easily figure out that if the clock period is less than 21ns, then either one or all the scenarios/cases/combinations fail.

2. So, we can easily conclude that for working of the entire circuit properly

    a. **Minimum clock period=Clock period(T3)=(Max data path delay)-(min clock path delay)+tsetup=26-9+4=21ns.**

    So, in general;

    Minimum clock period=(Max data path delay)-(min clock path delay)+tsetup

    And "Maximum Clock Frequency=1/(Min Clock Period)"

Example 3: Circuit with multiple Combinational paths between 2FFs:

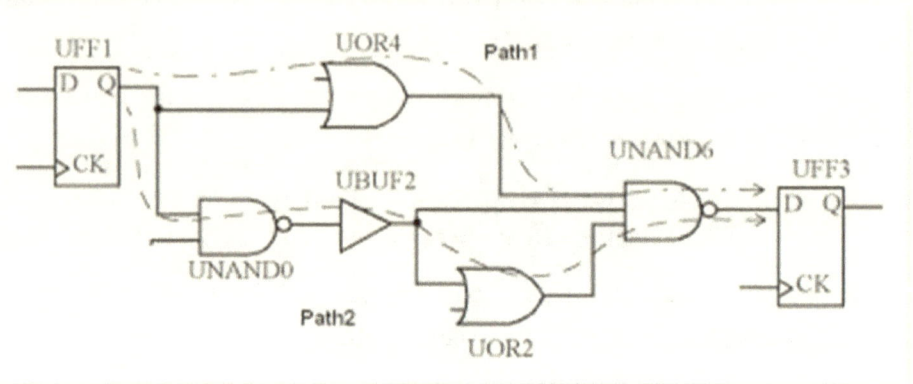

Figure 4.34:

Now the same scenario as with this example. I am not going to explain in detail. It's just like that if you have multiple paths in between the 2-flip flops, then as we have done in previous examples, please calculate the delays.

Then calculate the period and see which one is satisfying all the condition. Or directly I can say that we can calculate the Clock period based on the delay of that path which has a big number.

Min Clock Time Period=TClock-q(of UFF1)+max(delay of Ptah1,delay of Path2)+Tsetup(of UFF3).

Example 4: Circuit with different kind of Timing paths:

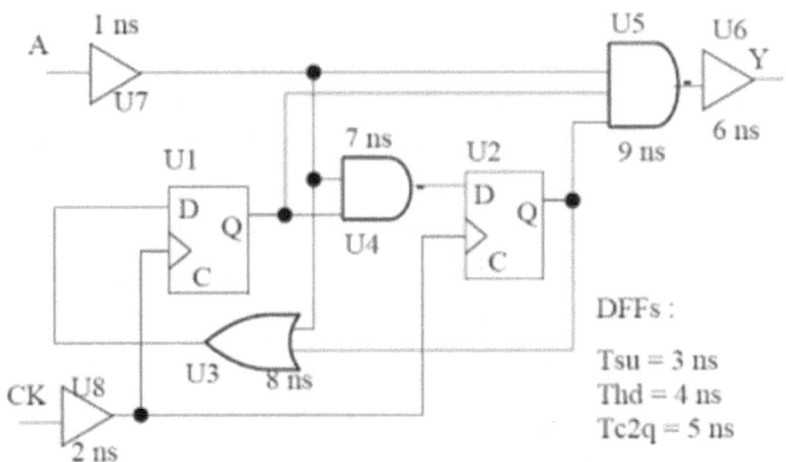

Figure 4.35:

You can easily figure out that in the above circuit, there are four types of data paths and two clock paths.

Data path:

1. Register to register Path
   a. U2->U3->U1(Delay=5+8=13ns)
   b. U1->U4->U2(Delay=5+7=12ns)
2. Input pin/port to Register(flip-flop)
   a. U7->U4->U2(Delay=1+7=8ns)
   b. U7->U3->U1(Delay=1+8=9ns)
3. Input pin/port to Output pin/port
   a. U7->U5->U6(Delay=1+9+6=16ns)
4. Register(flip-flop) to Output pin/port
   a. U1->U5->U6(Delay=5+9+6=20ns)
   b. U2->U5->U6(Delay=5+9+6=20ns)

Clock path:

1. U8->U1 (Delay=2ns)

2. U8->U2(Delay=2ns)

Now a few important points. This is not a full-chip circuit. In general, the recommendation is that you use registers at every input and output port. For the time being, we will discuss this circuit, considering this as a full-chip circuit. And you will how much analysis you have to do in this case. Next example, I will add the FF's (registers) at the input and output port, and then you come to know the difference.

Now let's study this circuit in more details

1. In this circuit, we must analyze in such a way that if we apply an input port at Port A, then how much time it will take to reach output Port Y. It will help us to find out the period of the clock.

2. Output pin Y relates to a three-input NAND gate. So, if we want a stable out at Y, we must make sure that all 3 Inputs of NAND gate should have stable data.

3. One input of NAND gate relates to Input pin A with the help of U7.
    a. Time taken by data to reach NAND gate is 1ns (gate delay of U7).

4. Second input pin of NAND gate relates to output pin Q of Flip flop U2.
    1. Time is taken by data which is present at input D of FF-U2 to reach NAND gate.
        a. 2ns (delay of U8)+5ns(Tc2q of FFU2)=7ns

5. Third input pin of NAND gate relates to the output pin Q of Flip Flop U1.
    1. Time is taken by data which is present at input D of FF-U2 to reach NAND gate.
        a. 2ns (delay of U8)+5ns(Tc2q of FF U1)=7ns

Note:

- I know you may have doubts about why the delay of U8 comes in picture.

  1. Concerning the clock edge at CLOCK pin, we can receive the data at NAND pin after 7ns only (Don't ask me why we can't take reference in the negative?)

- Maybe you can ask why we haven't considered the setup time of FF in this calculation.

  1. If in place of NAND gate, any FF would there we will consider the setup. We never consider the setup and Tc2q(Clock-2-Q) values of the same FF in the delay calculation at the same time. Because when we are considering Clock-2_Q delay, we assume that Data is already present at input Pin D of the FF.

  So, Time required for the data to transfer from input (A) to output(Y) pin is the maximum of

  Pin2Pin Delay=U7+U5+U6=1+9+6=16ns

  Clock2Out(through U1) delay=U8+U1+U5+U6=2+5+9+6==2ns

  Clock2Out(through U2) delay=U8+U2+U5+U6=2+5+9+6=2ns

  So out of this Clock2Out Delay is Maximum

From the above Study, you can conclude that data can be stable after 7ns at the NAND gate, and maximum delay is 22ns. And you can also assume that this much data is enough for calculating the Max Clock Frequency or Minimum Period. But that's not the case. Still, our analysis is half done in calculating the Max-clock-frequency.

As we have done in our previous example, we have to consider the path between 2 flip-flops also. So, the paths are:

- From U1 to U2 (Reg1Reg2)

  1. Path delay=2ns(Delay of U8)+5ns(TClock2Q of U1)+7ns(Delay of U4)+3ns(setup of U2)-2ns(Delay of U8)=17ns-2ns=15ns.

❖ From U2 to U1 (Reg2Reg1)
   1. Path delay=2ns(Delay of U8)+TClock2Q of U2 (5ns)+ Delay of U3 (8ns)+setup of U1 (3ns)-Delay of U8(2ns)=18ns-2ns=16ns

Note:

❖ I am sure you will ask why I subtracted "Delay of U8" from the above calculation :) because Delay of U8 is common to both the launch and capture path. So, we are not supposed to add this delay in our calculation. But to make it clear, I have added as per the previous logic and then subtracted it to make it clear.

So now if you want to calculate the maximum clock frequency, then you must consider all the delay which we have discussed above.

So

Max Clock Freq=1/Max(Reg1Reg2,Reg2Reg1,Clock2Out_1,Clock2Out_2,Pin2Pin)

=1/Max(15,16,22,22,16)

=1/22=45.5MHz

Example 5: Circuit with different kind of timing paths with the register at Input and output ports:

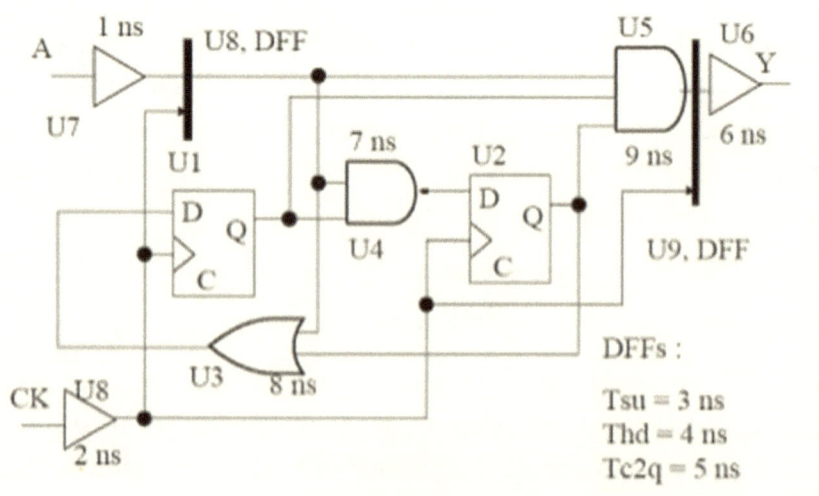

Figure 4.36:

In this example, we have just added 2 FFs U8 at Input pin and U9 at the output pin. Now for this circuit, if we want to calculate the max clock frequency, then it's like example1.

There are 7 Flip Flop to flipflop paths

1. U8->U4->U2
    a. Delay = 5ns+7ns+3ns=15ns
2. U8->U3->U1
    b. Delay =5ns+8ns+3ns=16ns
3. U8->U5->U9
    c. Delay=5ns+9ns+3ns=17ns
4. U1->U4->U2
    d. Delay=5ns+7ns+3ns=15ns
5. U1->U5->U9
    e. Delay=5ns+9ns+3ns=17ns
6. U2->U5->U9
    f. Delay=5ns+9ns+3ns=17ns
7. U2->U3->U1
    g. Delay=5ns+8ns+3ns=16ns

Since the maximum path delay is 17ns.

The minimum clock period for the circuit should be Tmin=17ns

And the maximum clock frequency is 1/Tmin=58.8 MHz.

Q19. Examples of Setup and Hold time?

Ans. **Problem 1: In the following Circuit, find out whether there is any Setup or Hold Violation?**

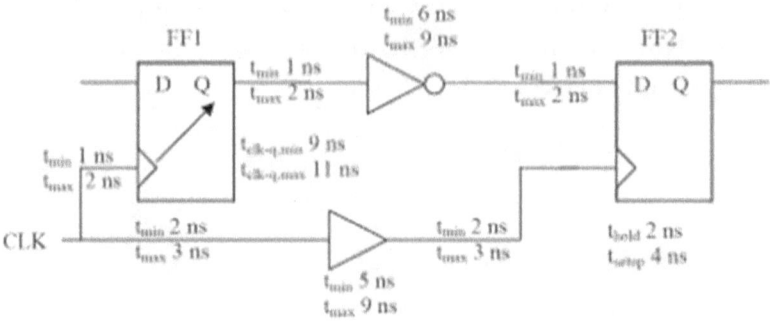

Figure 4.37:

Solution:

Hold Analysis:

When a hold check is performed, we must consider two things-

* Minimum Delay along the data path.
* Maximum Delay along the clock path.

If The difference between the data path and the clock path is negative, then a timing violation has occurred.

Data path is: CLOCK->FF1/CLOCK->FF1/Q->Inverter->FF2/D

Delay in Data path

=min(wire delay from CLOCK to Buffer input)+max(cell delay of Buffer)+max(wire delay from Buffer output to FF2/CLOCK pin)+(hold time of FF2)

=TClock=3+9+3+3=17ns.

Hold Slack=Td − TClock=18ns-17ns=1ns

Hold Slack is positive ->No hold Violation.

Note: If the hold time had been 4 ns instead of 2 ns, then there would have been a hold violation.

Td=18ns and TClock=3+9+3+4=19ns

So, Hold Slack=Td-TClock=18ns-19ns=-1ns (Violation)

Setup Analysis:

When a setup check is performed, we must consider two things-

- ❖ Maximum Delay along the data path.
- ❖ Minimum Delay along the clock path.

If the difference between the clock path and the data path is negative, then a timing violation has occurred.

Data path is: CLOCK->FF1/CLOCK->FF1/Q->Inverter->FF2/D

Delay in Data path

=max(wire delay of the clock input of FF1)+max(Clock-to-Q delay of FF1)+max(cell delay of inverter)+max(2 wire delay-"Qof FF1-to-inverter" and "inverter-to-D of FF2")

=Td=2+11+9+(2+2)=26ns

Clock path is:CLOCK->buffer->FF2/CLOCK

Clock path Delay

=(Clock period)+min(wire delay from CLOCK to Buffer input)+min(cell delay of Buffer)+min(wire delay from Buffer output to FF2/CLOCK pin)-(Setup time of FF2)

=tClock=15+2+5+2 – 4=20ns

Setup Slack=TClock-Td=20ns-26ns=-6ns

Since Setup Slack is negative->Setup violation

Note: A bigger clock period or a less maximum delay of the inverter solve this setup violations in the circuit.

E.g

If Clock period is 22ns then

TClock=22+5+2-4=31-4=27ns AND Td=26ns

Setup Slack=TClock-Td=27-26=1ns(No Violation)

Problem 2: To work correctly, what should be set-up and Hold Time at Input A in the following circuit. Also, find out the maximum operating frequency for this circuit. (Note: Ignore Wire delay). Where Tsu-Setup time; Thd-Hold Time; Tc2q-CLock-to-Q delay

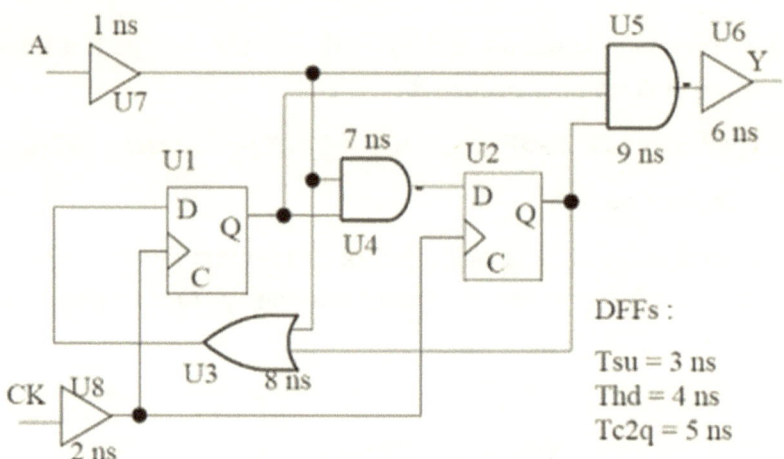

Figure 4.38:

Solution:

Step 1: Find out the maximum Register to register Delay.

Max Register to Register Delay

=(Clock-to-Q delay of U2)+(cell delay of U3)+(all wire delay)+(setup time of U1)

=5+8+2=16ns

Note:
- There are two register to register paths
  1. U2->U3->U1(Delay=5+8+3=16ns)
  2. U1->U4->U2(Delay=5+7+3=15ns)
- We must pick the maximum one

Step 2: Find Out Setup Time:

A setup time=Setup time of Flip Flop+Max(Data path Delay) – min(Clock path Delay)

=(Setup time of Flipflop+A2D max Delay)-(Clock path min delay)

=Tsu+(Tpd U7 + Tpd U3+wire delay)-Tpd U8

=3+(1+8) – 2=10ns

Note:

- Here we are not using Clock period. Because we are not supposed to calculate the Setup violation. We are calculating Setup time.
- All the wire delay is neglected. If Wire delay present, we must consider that one.
- There are 2 Data paths
    1. A->U7->U4->D of U2 (Data path Delay=1+7=8ns)
    2. A->U7->U3->D of U1(Data path Delay=1+8=9ns)
- Since for Setup calculation we need maximum Data path delay, we have chosen 2nd for our calculation

Step 3: Find out Hold Time:

Hold time=Hold time of Flip Flop + max(Clock path Delay) – min(Data path delay)

=(Hold time of Flip Flop+Clock path max delay) - (A2D max delay)

=Thd+Tpd U8 - (Tpd U7 + Tpd U4+wire delay)

=4+2-(1+7) = -2ns

Note: Same explanation as for the Setup time. For half time, we need a minimum data path, so we have picked the first Data path.

Step 4: Find out Clock to Out Time:

Clock to Out

=Cell delay of U8 + Clock-to-Q delay of Flip FLop + Cell delay of U5 + Cell delay of U6 + (all wire delay)

=Tpd U8 + U2 Tc2q + U5 Tpd +U6 Tpd

=2+5+9+6 = 22ns

Note:

- There are 2 Clock to Outpath-one from Flip flop U1 and other from U2.
- Since in this case the Clock-to-Q path for both Flip Flop us same, we can consider any path. But in some Circuit where the delay id different for both paths, we should consider Max delay path.

Step 5: Find Pin to Pin Combinational Delay (A to Y delay).

Pin to Pin Combination Delay (A to Y)

=U7 Tpd + U5 Tpd + U6 Tpd

=1+9+6=16ns

Step6: Find out Max Clock Frequency:

Max Clock Freq=1/Max(Reg2reg, Clock2Out, Pin2Pin)

=1/Max(16,22,16)

=45.5 MHz

So summary is:

Parameter	Description	Min	Max	Units
Tclk	Clock Period	22		ns
Fclk	Clock Frequency		45.5	MHz
Atsu	A setup time	10		ns
Athd	A hold time	-2		ns
A2Y	A to Y Tpd		16	ns
Ck2Y	Clock to Y Tpd		22	ns

Note: Negative hold times are typically specified as 0ns.

Problem 3: In the above Circuit, try to improve the timing by adding any "buffer" or "Register".

Solution:

The best way of doing this is "Register all Input and Output". We are adding DFF so same specification (as U2 and U1).

Now follow all those five steps on by one.

Step 1:

Max Register to Register Delay

U2 Tc2q + U5 Tpd + U9 Tsu = 5+9+3=17ns

Note:

- A lot of register to Register path
    1. U8->U5->U9 (Delay =5+9+3=17ns)
    2. U8->U4->U2 (Delay =5+7+3=15ns)
    3. U8->U3->U1(Delay=5+8+3=16ns)
    4. U1->U4->U2(Delay=5+7+3=15ns)
    5. U1->U5->U9(Delay=5+9+3=17ns)
    6. U2->U5->U9(Delay=5+9+3=17ns)
    7. U2->U3->U1(Delay=5+8+3=16ns)
- Maximum delay is 17ns, just picked anyone

Step 2:

A setup time = Tsu + A2D Tpd max - ClockTpd min

=Tsu + (Tpd U7)-Tpd U8

=3 + (1) -2 =2ns

Note: Only One path between A and D of FF (i.e U8)

Step 3:

A hold time = Thd + ClockTpd max - A2D Tpd min

= Thd + Tpd U8 -(Tpd U7)

= 4 - 2 - (1) = 5ns

Note: Only One path between A and D of FF (i.e. U8)

Step 4:

Clock to out:

= Tpd U8 + U9 Tc2q + U6 Tpd

= 2+5+6 = 13ns

Step 5:

No direct link between A and Y. So Not Applicable.

Step 6:

Max Clock Freq = 1/Max(Reg2reg, Clock2Out, Pin2Pin)

= 1/Max(17,13)

= 58.8 MHz

Parameter	Description	Min	Max	Units
Tclk	Clock Period	17		ns
Fclk	Clock Frequency		58.8	MHz
Atsu	A setup time	2		ns
Athd	A hold time	5		ns
Ck2Y	Clock to Y Tpd		13	ns

I hope this much will help you. Now it's time to summarize all the important things and formulas.

Points to remember:

1. Setup is checked at the next clock edge.
2. The hold is checked at the same clock edge.
3. For hold Check (Checking of hold Violation)
   - **Minimum Delay** along the **data path**.
   - **Maximum Delay** along the **clock path**.
4. For Setup Check (Checking of Setup Violation)
   1. **Maximum Delay** along the **data path**.
   2. **Minimum Delay** along the **clock path**.

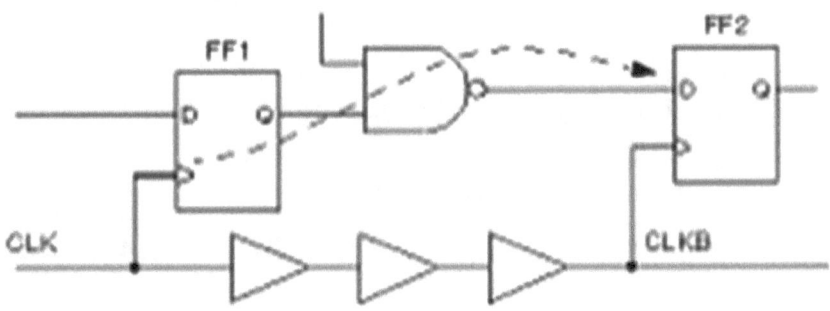

Figure 4.39:

Calculation of Setup Violation Check: Consider the above circuit of 2 FF connected.

Setup Slack=Required time - Arrival time (since we want data to arrive before it is required)

Where:
- Arrival time (max) = clock delay FF1(max) + clock-to-Q delay FF1 9max) + comb. Delay(max)

- ❖ Required time= clock adjust + clock delay FF2 (min) - setup time FF2
- ❖ Clock adjust= clock period (since setup is analyzed next edge)

Calculation of Hold Violation Check: Consider above circuit of 2 FF connected to each other.

Hold Slack = Arrival Time - Required time (since we want data to arrive after it is required)

Where:

- ❖ Arrival time(min)=clock delay FF1(min) + clock-to-Q delay FF1 (min) + comb. Delay(min)
- ❖ Required time = clock adjust + clock delay FF2 (max) + hold time FF2
- ❖ Clock adjust = 0 (since hold is analyzed at same edge)

Calculation of Maximum Clock Frequency:

Max Clock Freq=1/Max(Reg2reg delay, Clock2Out Delay, Pin2Pin delay)

Where:

- ❖ Reg2Reg Delay=Clock-to-Q delay of first FF (max) + comb. Delay (max) + setup time of 2ns FF
- ❖ Clock2Out Delay= Clock delay w.r.t FF 9max) + clock-to-Q delay of FF1 (max) + comb. Delay 9max)
- ❖ Pin2Pin delay= Comb delay between input pin to output pin(max)

Q20. How to solve Setup and Hold violations? (basic example)

Ans. **Basic of Fixing the "SETUP and HOLD" violations**

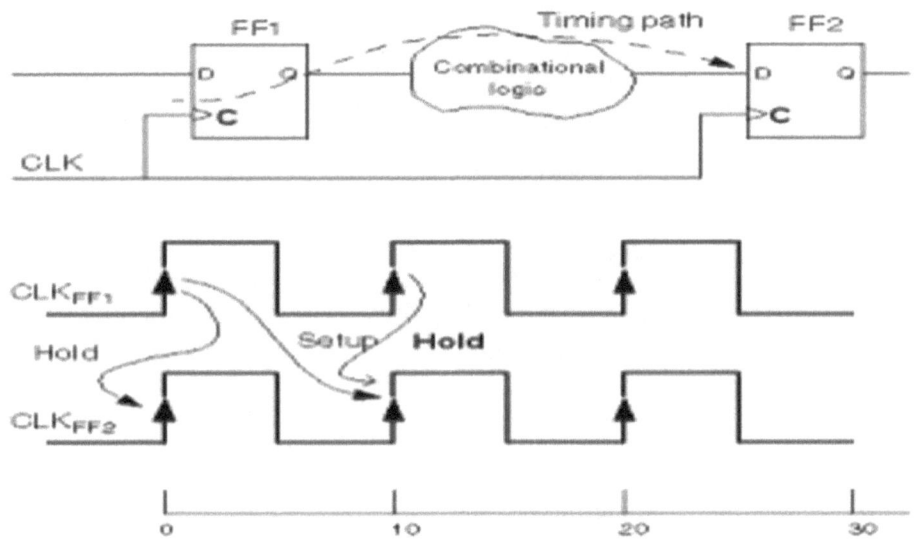

Figure 4.40:

Let's start with the following Diagram and consider this as common for the next few examples

In the following examples, we will pick different values of Setup/Hold values of Capture FF and Combinational Path Delay. Through these examples, we will study- how the setup and hold violations are dependent on each other and the delay of the circuit. If these things are clear then it's very easy for you to understand-- how can we fix the violations and if we are using any methods, then why?

Example 1:

Specification of the FF Circuit					
Setup	Hold	Clock period	Tck2q delay	Net Delay	Combinational Logic Delay
2ns	1ns	10ns	0ns (Ideal)	0ns (Ideal)	0.5ns

Let's discuss the flow of the data from FF1 to FF2

- ❖ Data is going to launch from FF1 at positive Clock Edge at 0 ns, and it will reach to Ff2 after 0.5ns (Combinational logic delay only)

- ❖ This data is going to capture at Ff2 at positive Clock Edge at 10ns.
- ❖ As per the Setup definition, data should be stable 2ns (Setup time of FF2) at FF2 before the positive Clock Edge (which is at 10ns)
  1. In the above case-data become stable 9.5 ns before the Clock edge at 10ns (10ns-0.5ns). This means it satisfies the Setup condition. **NO SETUP VIOLATION.**
- ❖ At the FF1-second set of data is going to launch at t=10ns and it will reach the Ff2 in another 0.5ns, means at t=10.5ns.
- ❖ This second set of data is going to update/override the first set of data.
- ❖ As per the Hold Definition, data should be stable till 1ns (Hold time of FF20 at FF2 after the clock edge (which is at 10ns)
  1. In the above case-first set of data is going to override by the second set of data at 10.5ns (means just after 0.5ns of the positive clock edge at FF2). This means it does not satisfy the hold condition. HOLD VIOLATION.

To find this Hold violation-we must increase the delay of the Data path so that the second set of data should not reach before t=11ns(10ns+1ns). That means the minimum delay of the Combination Logic Path to be 1 ns for NO HOLD VIOLATION. That means if you want to fix the HOLD violation, you can increase the Delay of the Data path by any method.

But it doesn't mean that you can increase the Delay by any Value. Let's assume that you have increased the delay of a combinational path by adding an extra buffer (with a delay of 8.5 ns). Now new specifications are

Specification of the FF Circuit					
Setup	Hold	Clock period	Tck2q delay	Net Delay	Combinational Logic Delay
2ns	1ns	10ns	0ns (Ideal)	0ns (Ideal)	=0.5ns + 8.5ns = 9ns

As pe the Setup definition, data should be stable 2ns (Setup time of FF2) before the Clock Edge (at Ff2 which is at 10ns) and with the updated specification -data will be stable at t=9ns, just 1ns before the clock edge at t=10ns at FF2. that means it does not satisfy the Setup condition. **SETUP VIOLATION.**

Since Data path delay is more than 1ns, there is **NO HOLD VIOLATION.**

So, it means that if we want to fix the setup violations, the Delay of the combinational path should not be more than 8ns (10ns-2ns). Means 8ns is the maximum value of the Delay of the Combinational Logic path forNO SETUP VIOLATION.

So, we can generalise this-

For NO HOLD and SETUP VIOLATION, the delay of the path should be in between 1 ns and 8 ns.

OR

For Violation free Circuit.;

Min delay of Combinational path>Hold time of Capture FF.

Max delay of Combinational path <Clock Period-Setup time of Capture FF.

Example2:

Specification of the FF Circuit					
Setup	Hold	Clock period	Tck2q delay	Net Delay	Combinational Logic Delay
6ns	5ns	10ns	0ns (Ideal)	0ns (Ideal)	0.5ns

The flow of the data from FF1 to FF2

- ❖ Data is going to launch from FF1 at Clock Edge at 0ns, and it will reach to FF2 after 0.5ns (combinational logic delay only).
- ❖ This data is going to capture at FF2 at Clock Edge at 10ns.
- ❖ **As per the Setup definition**, data should be stable 6ns (Setup time of FF2) before the Clock Edge (which is at 10ns)

1. In the above case-data become stable 9.5ns before the Clock edge at 10ns(10ns-0.5ns). That means it satisfies the Setup condition. **NO SETUP VIOLATION.**

❖ At the FF1-second set of data is going to launch at t=10ns and it will reach the FF1 in another 0.5ns, means at 10.5ns.

❖ This second set of data is going to update/override the first set of data.

❖ As per the Hold Definition, data should be stable till 5ns (Hold time of FF2) after the clock edge (which is at t=10ns) at FF2

1. In the above case-first set of data is going to override by the second set of data at 10.5 ns (means just after 0.5ns of the Clock edge at FF2). This means it does not satisfy the hold condition. HOLD VIOLATION.

To fix this Hold violation-(As per the previous example) we may increase the delay of the Data path so that the second set of data should not reach before t=15ns (10ns+5ns). That means the minimum delay of the Combination Logic Path should be 5 ns for NO HOLD VIOLATION.

But Now if you will verify the Setup condition once again (with combination delay of 5ns -which we have assumed for fixing the hold violation) then you come to know that data is going to stable only after 5ns (means 10ns-5ns=5ns before the clock edge at t=10ns). But as per the setup condition, data should be stable before 6ns. So, it means now it does not satisfy setup condition means SETUP VIOLATION.

So, in this scenario, we can't fix the setup and hold violations at the same time by adjusting the delay on the combinational logic.

You can also see it directly with the help of minimum and maximum value of combinational delay.

Min delay > Hold time of Capture FF (means 5ns)

Max Delay<Clock Period - Setup time of capture FF (Means 10ns - 6ns = 4ns)

So, Min delay > 5ns and Max Delay < 4ns which is not possible.

Now the point is how to fix these violations? This is a non-fixable issue until you change the clock frequency or replace the FF with lesser setup/hold value.

Let me explain this.

Min delay has dependence only on Hold time, which is fixed for a particular FF.

Max delay has a dependence on two parameters-Clock Period and Setup time-where Setup time is fixed for a particular FF. So, if you can change the FF with lower setup/hold violations, then you can fix this issue. But in case if that's not possible, then we must change the Clock period.

In case we are changing the Clock period:

Keep--Min delay>=5ns(No HOLD Violation)

Setup violation is by 6ns-5ns=1ns (6 ns= Setup time and 5ns= combinational delay). What if we will increase the Clock period by 1 ns. Means New clock period should be >11ns.

So far Clock Period 11ns:

Max delay<=Clock period (11ns)-Setup time(6ns)=5ns

Now-Max Delay=min Delay=5ns (Neither Hold nor Setup Violation)

We can generalise-

For Violation Free Circuit

Clock Period>=Setup time+Hold time.

Summary:

Min delay of Combinational path> Hold time to capture FF.

Max delay of Combinational path < Clock Period-Setup time of Capture FF.

Clock Period>=Setup time+Hold time.

Q21. How to solve Setup and Hold Violations? (Advance Example)

Ans: In this question, we will discuss a few more examples with more restrictions. Like

❖ What if we can't reduce the Delay of Data path?

Let's consider the following figure common to all examples

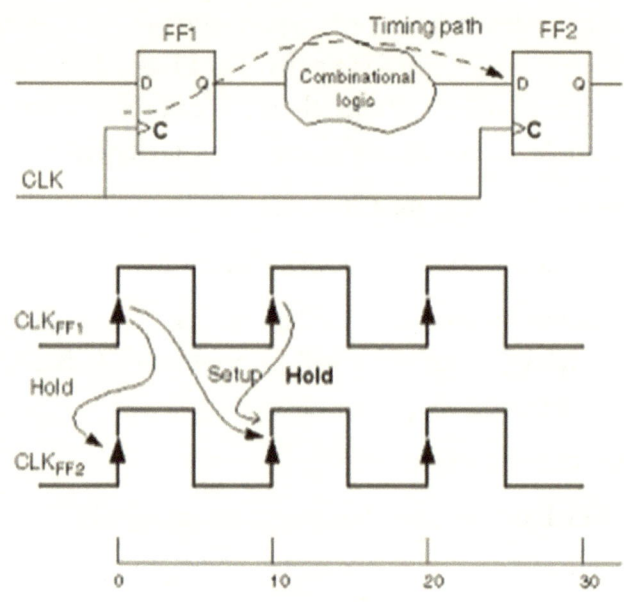

Figure 4.41:

Example3:

Specification of the FF Circuit					
Setup	Hold	Clock Period	Tck2q delay	Net Delay	Combinational Logic Delay
3ns	2ns	10ns	0ns(Ideal)	0ns(Ideal)	5ns (can't be further reduced)

Clock period Condition: (Satisfied)

Setup time + Hold time = 5ns

Clock period=10ns

Clock Period > Setup time + Hold time (10>5)

Min delay / Hold Condition: (Satisfied)

Combinational Delay (5ns)>Hold time.

Means- NO HOLD VIOLATION

Max Delay/Setup Condition: (Satisfied)

Combinational delay 95ns)<Clock period (10ns)-Setup(3ns)

Means- NO SETUP VIOLATION

Example 4:

Specification of the FF Circuit					
Setup	Hold	Clock Period	Tck2q delay	Net Delay	Combinational Logic Delay
4ns	3ns	10ns	0ns (Ideal)	0ns (Ideal)	8ns (can't be further reduced)

Clock Period Condition: (Satisfied)

Setup time + Hold time = 4ns+3ns=7ns

Clock period = 10ns

Clock Period > Setup time + Hold time (10>7)

Min delay / Hold Condition: (Satisfied)

Combinational Delay (8ns)>Hold time(3ns)

Means- NO HOLD VIOLATION

Max Delay/Setup Condition: (Not Satisfied)

Combinational delay (8ns) **Is Not Less Than** "Clock period (10ns)-Setup (4ns)"

Means- SETUP VIOLATION

Since we can't change this Combinational delay and Setup time for the FF, so we must think of something else. Since we can't touch the data path, we can try with the clock path.

The flow of the data from FF1 to FF2:

- ❖ Let's assume that you have added one buffer of Tcapture delay in the clock path between FF1 and FF2.
- ❖ Data is going to launch from FF1 at Clock Edge at 0ns, and it will reach to FF2 after 8ns (combinational logic delay only).
- ❖ This data is going to capture at FF2 at Clock Edge at 10ns+T_capture(because of Delay added by Buffer).
- ❖ **As per the Setup definition**, data should be stable 4ns (Setup time of FF2) before the Clock Edge at FF2 and in the above case clock edge is at t=T_capture +10ns.

So, for NO Setup violation

=>8ns(Combinational Delay)<t_capture + 10ns (clock period)-4ns(Setup Time of FF2)

=>12ns-10ns<T_capture

=>T_capture>2ns

Let's assume if my T_capture=3ns. Then NO SETUP VIOLATION.

Now, recheck the Hold violation

- ❖ At the FF1-second set of data is going to launch at t=10ns and it will reach the FF2 in another 8ns, means at t=18ns.
- ❖ This second set of data is going to update/override the first set of data present at FF2.
- ❖ As per the Hold definition, data should be stable till 3ns (Hold time of FF2) after the clock edge at FF2 (which is at t=10ns + 3ns=13ns- where 3ns is the T_capture).

❖ That means Data should be remain stable till= 13ns + 3ns=16ns.

1. In the above case, the second set of data is going to override only after t=18ns. That means the first set of data remain stable until 16ns means NO HOLD VIOLATION.

Let me Generalize this concept.

I am sure; few people may have a question that what will happen if we will add the buffer in the Launch path. Let's discuss that. Please consider the following diagram for this. In his Launch Clock path has a buffer with a delay of "T-launch" and capture clock path has another buffer of delay "T_capture".

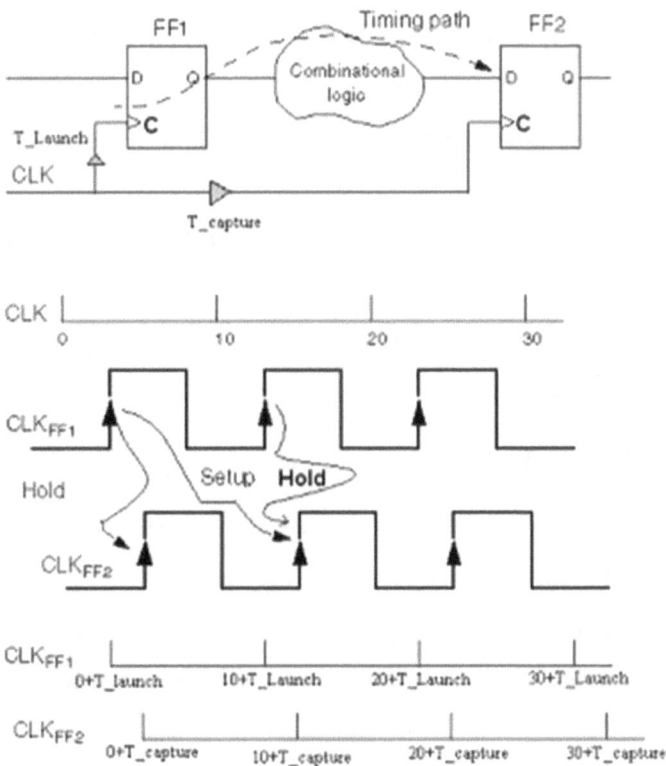

Figure 4.42:

SPECIFICATION OF THE FF CIRCUIT	
Setup	T_setup
Hold	T_hold
Clock Period	Clock_period
Tck2q Delay	0(Ideal)
Net Delay	0(Ideal)
Combinational Logic Delay (b/w 2FF's)	Td
Launch Clock Path Delay	T_launch
Capture Clock Path Delay	T_capture

Let's understand the data flow from FF1 to FF2

- Data is going to launch from FF1 at Clock Edge at T-launch, and it will reach to FF2 after Td(combinational logic delay only) that means t=Td + T_launch.

- This data is going to capture at FF2 at Clock Edge at "Clock_period + T_capture".

- **As per the Setup definition,** data should be stable "t_setup"(Setup time of FF2) time before the Clock Edge at FF2

    1. Means data should reach at FF2 before t="Clock_period + t_capture -T_setup".

So, for NO SETUP VIOLATION:

=>T_launch +Td<Clock_period + t_capture -T_setup

=>**Td<Clock_Period + (T_capture - T_launch) - T_setup**

- A the FF1-second set of data is going to launch at t= "Clock_period + T_launch" and it will reach the FF2 in another Td, means at t=Clock_Period + Td + T_launch.

- This second set of data is going to update/override the first set of data present at FF2.

❖ **As per the Hold Definition**, data should be stable till "T_hold"(Hold time of FF2) time after the Clock edge (which is at t= "Clock_period + T_capture").

1. Means Next set of data should not reach FF2 before t= "Clock_period + T_capture + T_hold"

So, for NO HOLD VIOLATION:

=>Clock_period + Td + T_launch>Clock_Period + T_capture + T_hold

=>Td>(T_capture - T_launch) +T_hold

Summary for this Answer:

Clock Period Condition:

Clock period > Setup time + Hold Time

Max Delay/Setup Condition:

Td<Clock_Period + (T_capture - T_launch) - T_setup

Min Delay/Hold Condition:

Td>(T_capture - T_launch) + T_hold

Q22. How to solve Setup and Hold Violations? (More Advance Examples)?

Ans. In the last question, we have discussed two more examples with different specifications with more restrictions. (Both net delay and Tck2Q wire ideal means 0ns) and figure out that if you want to fix the violation by increasing/decreasing delay in the data path, then the following condition should be satisfied.

Min delay of Combinational path > hold time of capture FF.

Max delay of Combinational path < Clock Period - Setup time of Capture FF.

Clock Period >= Setup time + Hold time.

But in case if you can't touch the data path and you have to increase/decrease the delay in the clock path (means between "Clock pin to Launch FF clock pin: or between " Clock pin and capture FF clock pin"), the following conditions should be satisfied.

Max Delay/Setup Condition:

Td<Clock_Period + (T_capture - T_launch) - t_setup

Min Delay / Hold Condition:

Td>(T_capture - T_launch) + T_hold

Where:

Td->Combinational path delay (between the 2 FFs)

T_capture-> Delay of circuit present between "Clock pin and capture FF clock pin"

T_launch->Delay of circuit present between "Clock pin to Launch FF clock pin"

In this question, we will discuss a few more examples with more restrictions.

Let's consider the following figure common to all examples until unless it's specified.

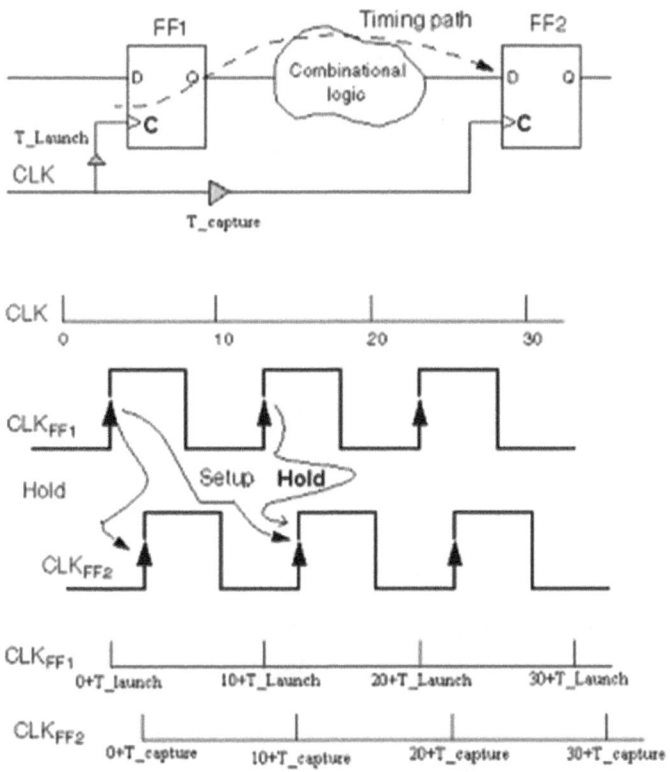

Figure 4.43:

Example 5:

Specification of the FF Circuit					
Setup	Hold	Clock Period	Tck2q delay	Net Delay	Combinational Logic Delay
3ns	2ns	10ns	0ns (Ideal)	0ns (Ideal)	11ns (can't be further processed)

Clock Period Condition: (Satisfied)

Setup time + Hold times = 5ns

Clock period=10ns

Clock Period > Setup time + Hold time (10>5)

Min delay/hold Condition:(Satisfied)

Combinational Delay (11ns)>Hold time.

Means-NO HOLD VIOLATION

Max Delay/Setup Condition:

Combinational delay (11ns) **Is Not Less than** "Clock period(10ns)-Setup(3ns)"

Means -SETUP VIOLATION.

Since adding a delay in the data, the path is not going to fix this violation, and we can't reduce the combinational delay. So, as we have discussed in our previous question, we will try with the clock path.

From the last question, if $T\_capture$ is the delay of the buffer which is inserted between the CLOCK and Capture's FF and T_launch is the delay of the buffer which is inserted between the CLOCK and Launch's FF, then

Max Delay/Setup condition is:

Td<Clock Period + (T_capture -T_launch)-T_setup

=>11ns<10ns-3ns+(T+capture- T_launch)

=>11ns<7ns+(T_capture - T_launch)

=>4ns<(T_capture - T_launch)

Now we can choose any combination of T_capture and T_launch such their difference should be less than 4ns.

Note: Remember in the design if you are fixing the violation by increasing or decreasing the delay in the Clock path then always prefer not to play too much with this path.

I never prefer to use T_launch in this case (For setup fixing, I ignore to use T_launch).

So, let's assume T_launch=0ns and T_capture=5ns.

Then

11ns<7ns+5ns **means no Setup Violation.**

Check once again the Hold condition.

Min delay/ Hold Condition:

Td>(T_capture - T_launch) + T_hold

=>11ns >(T_capture - T_launch)+T_hold

=>11ns>5ns+2ns

=>11ns>7ns **means NO HOLD VIOLATION**

Example 6:

Specification of the FF Circuit					
Setup	Hold	Clock Period	Tck2q delay	Net Delay	Combinational Logic Delay
3ns	5ns	10ns	0ns (Ideal)	0ns (Ideal)	2ns (can't be further reduced and we can't increase the delay in the data path by any methods)

Let's check the conditions directly

Clock Period Condition (Satisfied):

Setup time + Hold time = 8ns

Clock period = 10ns

Clock Period>Setup time + Hold time (10ns>8ns)

Means we can fix violations if there is any.

Max Delay/Setup Condition(Satisfied):

Td<Clock_Period + (T_capture - T_launch)-T_setup

Combinational Delay=2ns

There is no delay in the clock path till now, so T_capture= T_launch=0ns

=>Td(2ns) <Clock_period (10ns) +0ns - T_setup (3ns)

=>2ns < 7ns **means NO SETUP VIOLATIONS**

Min Delay/Hold Condition(Not Satisfied):

Td>(T_capture - T_launch) + T_hold

Combinational Delay = 2ns

There is no delay in the clock path till now, so t_capture = T_launch = 0ns

=>Td(2ns) is not greater than 0ns + T_hold (5ns)

Means HOLD VIOLATION

Since we can't make change in the delay path, so we must touch the clock path.

For Hold fixing-

=>Td>(T_capture- T_launch)+T_hold

=>2ns>(T_capture - T_launch)+5ns

=>-3ns(T_capture - T_launch)

For satisfying the above equation T_launch should have more value in comparison to T_capture.

We can choose any combination of T_capture and T_launch

Note: Remember in the design if you are fixing the violation by increasing or decreasing the delay in the Clock path then always prefer not to play too much with this path.

I will never prefer to use T_capture in this case (For Hold fixing, I ignore to use T_capture).

So, let's assume T_capture =0ns and T_launch =4ns

Then

T_launch+Td>5ns(hold time)

=>4ns+2ns>5ns **NO HOLD Violation.**

Check once again the Setup Condition:

Td<Clock Period +(T_capture - T_launch)-T_setup

=>2ns<10ns+0ns-4ns-3ns

=>2ns<3ns **means No Setup Violation**

Note:(T_capture - T_launch) also known as CLOCK SKEW.

Example 7:

Specification of the FF Circuit					
Setup	Hold	Clock period	Tck2q delay	Net Delay	Combinational Logic Delay
6ns	5ns	10ns	0ns(Ideal)	0ns(Ideal)	0.5ns

Clock Period Condition(Not Satisfied):

Setup time+Hold time=11ns

Clock period=10ns

Clock Period **is not greater than** Setup time + Hold time

Means we can't fix violations if there is any.

But still, we will try once again with all other conditions, to prove that the condition mentioned above should be valid for fixing the violations.

Max Delay/Setup Condition(Satisfied):

Td<Clock_Period + (T_capture - T_launch)-T_setup

Combinational Delay =0.5ns

There is no delay in the clock path till now, so T_capture=T_launch=0ns

=>Td(0.5ns)<Clock_period (10ns) +0ns-T_setup(6ns)

=>0.5ns<4ns-**Means NO SETUP Violations**

Min Delay / hold Condition (Not Satisfied):

Td>(T_capture - T_launch)+T_hold

Combinational Delay=0.5ns

There is no delay in the clock path till now, so T_capture= T_launch=0ns

=>Td(0.5ns) is not greater than 0ns + T_hold(5ns)

Means HOLD VIOLATION

If you want to fix the Hold Violation, then we already seen that by increasing/decreasing the delay the data path it can't be fixed. Even if this will be fixed, then Setup violation will occur.

Let's try with T_capture or T_launch. Means by adding a delay in the clock circuit.

As per the above equations/conditions and corresponding values:

Max Delay/Setup Condition:

Td<Clock Period + (T_capture - T_launch) - T_setup

=>Td<10ns-6ns+(T_capture - T_launch)

=>Td<4ns+(T_capture - T_launch)

Min Delay/ hold Condition:

Td>(T_capture - T_launch)+T_hold

=>Td >(T_capture - T_launch)+5ns

Remember all 3 variables Td, T_capture, T_launch is positive number.

Possible values of (T_capture - T_launch)=+/-A(where A is a positive number)

Case 1: (T_capture - T_launch)=+A

=>Td < 4ns +A - Condition(a)

=>Td > 5ns +A - Condition(b)

Satisfying both the conditions ("a" and "b") not possible for any positive value of A.

Case 1: (T_capture - T_launch)= -A

=>Td <4ns -A =>Td+A<4ns - Condition(a)

=>Td>5ns-A =>Td+A >5ns - Condition(a)

Satisfying both the conditions ("a" and "b") not possible for any positive value of A.

That means I am successfully able to prove that if the following condition is not satisfied, then you can't fix any violation by increasing/decreasing delay in either data_path or clock_path.

Clock Period > Setup time + Hold time.

Summary of this Question:

Clock Period Condition:

Clock period > Setup time + Hold Time

For fixing any violation (without changing the Clock period) - This condition should be satisfied.

Max Delay / Setup Condition:

Td <Clock_Period + (T_capture - T_launch) - T_setup

For Fixing the setup Violation - Always prefer T_capture over T_launch

Min Delay / Hold Condition:

Td > (T_capture - T_launch)+T_hold

For Fixing the hold Violation - Always prefer T_launch over T_capture.

PART V

IMPORTANT CODES

Q1. Write and Verilog code and Test Bench for Synchronous FIFO?

Ans.
```
module FIFObuffer (Clock,datain,rd,wr,en,dataout,rst, empty, full);
input Clock,rd, wr,en,rst ;
output empty,full;
input [31:0] datain;
output reg [31:0] dataout;
reg [2:0] count=0;
reg [31:0] FIFO [0:7];
reg [2:0] readcounter=0,writecounter=0;
assign empty =(count==0)?1'b1:1'b0;
assign full=(count==8)?1'b1:1'b0;

always @ (posedgeClock)
begin
if(en==0)
else begin
if(rst) begin
readcounter=0;
writecounter=0;
end
else if (rd==1'b1 &&count!=0) begin
dataout=FIFO[readcounter];
readcounter=readcounter+1;
end
else if(wr==1'b1 && count!=8) begin
```

```
FIFO[writecounter]=datain;
writecounter=writecounter+1;
End
else
end
if(writecounter==8)
writecounter=0;
else if(readcounter==8)
readcounter=0;
else;
if(readcounter>writecounter) begin
count=readcounter-writecounter;
end
else if (writecounter>readcounter)
count=writecounter-readcounter;
else;
end
endmodule

module fifobuffer_tb;
reg Clock,rd,wr, en,rst;
reg [31:0]datain;
wire [31:0]dataout;
wire empty, full;
fifouut (Clock(Clock),.datain(datain),.rd(rd),.wr(wr),.en(en),.dataout(dataout),.rst(rst),.empty(empty),.full(full));
initial begin
```

```
Clock=1'b0;
datain=32'h0;
rd=1'b0;
wr=1'b0;
en=1'b0;
rst=1'b1;
#100;
en=1'b1;
rst=1'b1;
#20;
st=1'b0;
wr=1'b1;
datain=32'h0;
#20
datain=32'h1;
#20;
datain=32'h2;
#20;
datain=32'h3;
#20;
datain=32'h4;
#20;
wr=1'b0;
rd=1'b0;
end
always #1- Clock=~Clock;
endmodule
```

Q2. Write verilog code and test bench for Asynchronous Up Counter?

Ans. Asynchronous means in terms of simple definition without external clock synchronization.

The output always remains free from the clock signal.

Generally, the first FF is clocked with the main external clock, and each of the next FF has an output of previous FF as their clock. This helps in reducing the number of FFs, and additional gates hence require less complexity.

It is now coming to the special "MOD" term. It stands for modulus.

When you must design a Mod-Y counter, then the basic steps include

1. The equation -: $2^x = Y$.

2. Now find the value of X if you know basic Maths. You can use logarithms

Thus, after getting the value of X, you get how many FFs are required hence you require X FFs to design Mod- Y UP or Down counter.

Here is the block diagram of Mod-16 or 4bit Asynchronous UpCounter

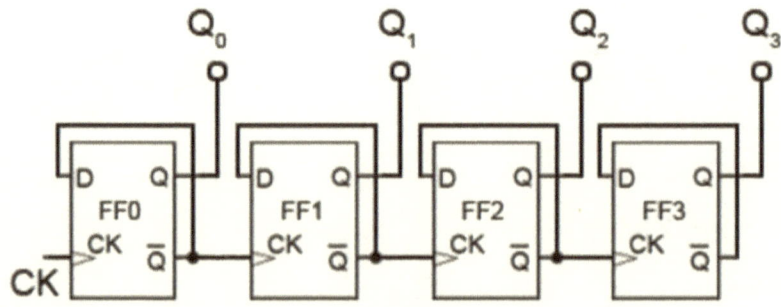

Figure 5.1:

Now for Mod-16 we have value of X as 4 hence 4 FFs

Here is the code to test this

module dff(d,Clock, q);

```verilog
input d;
input Clock;
wire d;
wire Clock;
output reg q;
initial
q=1'b0;
always @ (posedgeClock) begin
q<=d;
end
endmodule
module fourbitup();
reg Clock;
reg d;
wire [3:0]q;
initial
Clock=0;
always #1 Clock=~Clock;
dff d1 (!q[0],Clock,q[0]);
dff d2 (!q[1],!q[0],q[1]);
dff d3 (!q[2],!q[1],q[2]);
dff d4 (!q[3],!q[2],q[3]);
endmodule
```

The block diagram for 4 bit Down Asynchronous Counter

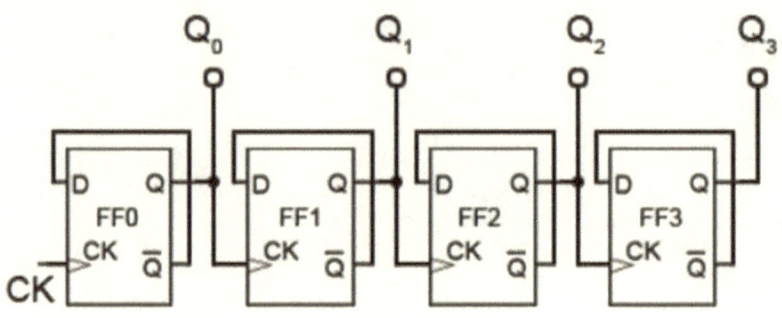

Figure 5.2:

Notice the clock inputs to each FF after 1st FF.

Here is the code for Down Counter 4 bit

module dff(d,Clock, q);

input d;

input Clock;

wire d;

wire Clock;

output reg q;

initial

q=1'b0;

always @ (posedgeClock) begin

q<=d;

end

endmodule

module fourbitup();

reg Clock;

```
reg d;
wire [3:0]q;
initial
Clock=0;
always #1 Clock=~Clock;
dff d1 (!q[0],Clock,q[0]);
dff d2 (!q[1],q[0],q[1]]);
dff d3 (!q[2],q[1],q[2]);
dff d4 (!q[3],q[2],q[3]);
endmodule
```

Q3. Write a verilog code and test bench for parity bit checker using function?

Ans:
```
module func();
reg [7:0] data;
reg parity;
integer i;
function abc;
input [31:0]data;
integer i;
begin
abc=0;
for(i=0 ;i<32 ; i=i+1) begin
abc=abc^data[i];
end
```

```
end
endfunction
initial
begin
parity=0;
data=0;
for(i=145 ;i<160 ; i=i+1)
begin
#5 data=i;
parity =abc(data);
$display("Data=%b,Parity_bit=%b",data,parity);
end
#25 $finish;
end
endmodule
```

Output

```
VSIM 13> run
# DATA = 10010001, Parity-bit = 1
# DATA = 10010010, Parity-bit = 1
# DATA = 10010011, Parity-bit = 0
# DATA = 10010100, Parity-bit = 1
# DATA = 10010101, Parity-bit = 0
# DATA = 10010110, Parity-bit = 0
# DATA = 10010111, Parity-bit = 1
# DATA = 10011000, Parity-bit = 1
# DATA = 10011001, Parity-bit = 0
# DATA = 10011010, Parity-bit = 0
# DATA = 10011011, Parity-bit = 1
# DATA = 10011100, Parity-bit = 0
# DATA = 10011101, Parity-bit = 1
# DATA = 10011110, Parity-bit = 1
# DATA = 10011111, Parity-bit = 0
```

Q4. Write a Verilog code for 2-bit Vedic multiplier?

Ans.

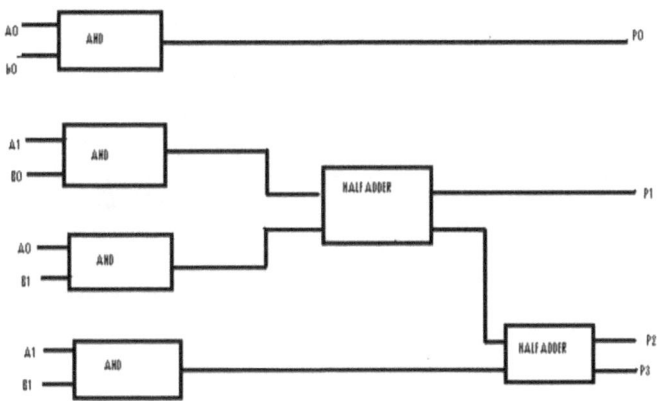

Figure 5.3:

module vedic_2bit_multiplier (a,b, c);

input [1:0]a;

input [1:0]b;

output [3:0]c;

wire [3:0]c;

wire [3:0]temp;

assign c[0]=a[0]&b[0];

assign temp[0]=a[1]&b[0];

assign temp[1]=a[0]&b[1];

assign temp[2]=a[1]&b[1];

ha ha1(temp[0],temp[1],c[1],temp[3]);

ha ha2(temp[2],temp[3],c[2],c[3]);

Endmodule

```
module ha(a,b,s,c);
input a,b;
output s, c;
assign s=a^b;
assign c=a&b;
endmodule
```

www.ingramcontent.com/pod-product-compliance
Lightning Source LLC
Chambersburg PA
CBHW030937180526
45163CB00002B/599